SpringerBriefs in Electrical and Computer Engineering

For further volumes:
http://www.springer.com/series/10059

Mohammad Rostami

Compressed Sensing with Side Information on the Feasible Region

 Springer

Mohammad Rostami
Singapore University of Technology and Design
Singapore
Singapore

ISSN 2191-8112   ISSN 2191-8120   (electronic)
ISBN 978-3-319-00365-8   ISBN 978-3-319-00366-5   (eBook)
DOI 10.1007/978-3-319-00366-5
Springer Cham Heidelberg New York Dordrecht London

Library of Congress Control Number: 2013937186

Printed on acid-free paper

Springer is part of Springer Science+Business Media (www.springer.com)

To my father and my mother

Preface

Reconstruction of continuous signals from a number of their discrete samples is central to digital signal processing. Digital devices can only process discrete data and thus processing the continuous signals requires discretization. After discretization, possibility of unique reconstruction of the source signals from their samples is crucial. The classical sampling theory provides bounds on the sampling rate for unique source reconstruction, known as the Nyquist sampling rate. Recently a new sampling scheme, Compressive Sensing (CS), has been formulated for sparse signals. CS is an active area of research in signal processing. It has revolutionized the classical sampling theorems and has provided a new scheme to sample and reconstruct sparse signals uniquely, below Nyquist sampling rates. A signal is called (approximately) sparse when a relatively large number of its elements are (approximately) equal to zero. For the class of sparse signals, sparsity can be viewed as prior information about the source signal. CS has found numerous applications and has improved some image acquisition devices.

Interesting instances of CS can happen, when apart from sparsity, side information is available about the source signals. The side information can be about the source structure, distribution, etc. Such cases can be viewed as extensions of the classical CS. In such cases we are interested in incorporating the side information to either improve the quality of the source reconstruction or decrease the number of the required samples for accurate reconstruction.

A general CS problem can be transformed to an equivalent optimization problem. We study a special case of CS with side information about the feasible region of the equivalent optimization problem. Based on spherical section property it is shown that the solution to such problems is unique and stable towards noise. In addition, an efficient algorithms is provided, which incorporates the side information to solve for better estimation of the source signal. Experiments confirm that the proposed algorithm converges to the solution faster and results in a more accurate estimation. Moreover, it is robust towards the noise power and also the noise model.

The proposed scheme is applied to practical problems: image deblurring in optical imaging, 3D surface reconstruction, and reconstructing spatiotemporally correlated sources. Experimental results confirm the usefulness and effectiveness of the proposed scheme. The results indicate that we can apply the proposed algorithm to improve the sampling devices in such applications without improving the hardware. Consequently, we can consider this tool as a low-cost technique compared to hardware improvement.

Mohammad Rostami

Acknowledgments

A major component of this book is based on my research during the MASc period in the University of Waterloo. I would like to express my appreciation to my supervisors, Professors Zhou Wang and Oleg Michailovich, who have guided and inspired me through every step during the course of my studies at University of Waterloo.

Some material of the book is derived from the research during my stay at Singapore University of Technology and Design (SUTD). I would like to appreciate Prof. Ngai-Man Cheung for his support and assistance during that period.

Last but not the least, my deep gratitude goes to my family members, especially my parents for their endless love and support.

Contents

Acronyms

ADC	Analog to Digital Convertors
AO	Adaptive Optics
ASF	Amplitude Spread Function
BP	Basis Pursuit
BP	Belief Propagation
BPDN	Basis Pursuit De-Noising
CCS	Conventional Compressed Sensing
CS	Compressed Sensing
DCS	Derivative Compressed Sensing
DS	Dense Sampling
GMA	Generalized Message Passing Algorithm
GPF	Generalized Pupil Function
IHT	Iterative Hard Thresholding
LS	Least Square
MGF	Moment Generating Function
MRI	Magnetic Resonance Imaging
MSE	Mean Squared Errors
OMP	Orthogonal Matching Pursuit
PS	Photometric Stereo
PSF	Point Spread Function
RIP	Restricted Isometry Property
SFS	Shape From Shading
SHI	Shack-Hartmann Interferometer
SNR	Signal to Noise Ratio
SSIM	Structural Similarity Index
SSP	Spherical Section Property
TV	Total Variation

Chapter 1
Introduction

Central to digital signal processing is the Shannon-Nyquist sampling theorem [1], which provides conditions under which a band-limited signal can be reconstructed via its discrete time samples uniquely. It states that the minimum uniform sampling rate for exact reconstruction of these signals is twice the signal band-width (Nyquist rate) in Fourier domain. This theorem has been extended for the case of bandpass [2] and random sampling [3]. This classical conclusion is crucial in signal processing and permits using digital devices to process natural continuous signals. Obviously we are interested in decreasing the sampling rate to reduce complexity, but according to the Shannon-Nyquist sampling theorem we are limited by the Nyquist rate in the general case. But it is easy to build special cases where signal reconstruction is possible with sampling rates less than the Nyquist rate. The question is, is it possible to identify such classes of signals?

Over the time the problem of reconstruction of signals subject to prior information gained interest among researchers. Is it possible to use the prior information to decrease the sampling rate? For instance, consider a very simple case where we know our signal of interest is a sinusoid $y(t) = A\sin(\omega t)$; it is trivial that in this extreme case one can reconstruct the signal via two non-zero samples to solve for A and ω. Whereas, the classical theorem suggests to use infinitely many samples with the sampling rate of $\frac{\pi}{\omega}$. It is easy to show that more generally any finite mixture of sinusoids can be reconstructed using finite number of samples (subject to knowing the number of sinusoids) [4]. This simple example confirms how prior information can be used to decrease the classical sampling rate, needed for reconstruction. Although the mixture of sinusoids model has practical importance but such applications are limited. Do we have a wider class of signals with similar property?

In around 2005, a new sampling scheme was formulated for sparse signals, i.e., signals which admit a sparse representation in a predefined basis/frame. This theory, nowadays known as compressed sensing (aka compressive sampling), asserts that sparse signals can be recovered from their discrete measurements, whose number is proportional to the ℓ_0-norm of the coefficients of the sparse representation. As a result, cases are numerous in which the sampling efficiency of compressed sensing (CS) far

M. Rostami, *Compressed Sensing with Side Information on the Feasible Region*,
SpringerBriefs in Electrical and Computer Engineering,
DOI: 10.1007/978-3-319-00366-5_1, © The Author(s) 2013

supersedes that of the classical Shannon-Nyquist sampling [5, 6]. Over the last decade many researchers have worked on this topic and developed many interesting results. The results have been used in some areas of signal processing and communications.

1.1 Compressed Sensing

Compressed Sensing has been formulated and studied mathematically in [5–7] and later became a major subject of interest. The theory is based on sparse prior assumption on the source in the sampling problem. It is interesting to note that prior to development of this theory, its reconstruction method had been studied fairly well. In fact prior algorithms for convex optimization of l_1-norm made CS a viable technique.

Definition 1 *Assume $x \in \mathbb{R}^n$ is a discrete finite signal. It is called k-sparse if its representation, s, in basis $\Psi \in \mathbb{R}^{n \times n}$ has at most k nonzero elements:*

$$x = \Psi s \tag{1.1}$$

also we define support of a vector s as: $supp(s) = \{i \mid s_i \neq 0\}$,

similarly we call the source approximately sparse if at most k elements of the source are greater than a (small) threshold that we set, i.e., if we sort the signal values, the signal values decay rapidly after the kth element. For instance natural images have this property in DCT domain. It is trivial that a k-sparse signal can be stored using $2k$ numbers (Only we need to store the value and the index of k nonzero element). This property has been used to design compression algorithms such as JPEG for natural images or MPEG for video streams.

This class of signals were known before the development of CS theory and were called compressible signals in the literature. CS theory provides conditions under which one can sense this class of signals compressibly. In CS framework we assume that we use non-adaptive linear sampling, i.e.:

$$\mathbf{y} = \Phi \Psi \mathbf{s}, \tag{1.2}$$

where $\mathbf{y} \in \mathbb{R}^m$ is vector of samples and $\Phi \in \mathbb{R}^{m \times n}$, $m < n$ is a full rank matrix, called sensing matrix. This means that instead of point sampling, in each measurement we measure a linear combination of signal samples. The value of m specifies the sampling rate. To reconstruct the source signal $\mathbf{s}$ the undetermined system (1.2) must be solved. In the general case we have infinitely many solutions ($m < n$) with a feasible region automorphic to null(Ψ) $\equiv \mathbb{R}^{n-m}$. But k-sparsity condition limits the feasible region and unique solution might be expected if the signal is sparse enough. Two main questions are:

1. Under what condition (1.2) has a unique solution (given Φ, Ψ, $\mathbf{y}$)?
2. In case of a unique solution how one can solve (1.2) for that unique solution?

Let's focus on the matrix $A = \Phi\Psi$. Uniqueness of the solution for (1.2) holds if no pairs of two distinct k-sparse signals result in the same samples: $A\mathbf{s}_1 = A\mathbf{s}_2 = \mathbf{y} \rightarrow \mathbf{s}_1 = \mathbf{s}_2$. This means that the difference of no two distinct k-sparse signals must not lay in nullspace of A. So linear independence of any $2k$ combinations of columns of A is a necessary condition for uniqueness. Vandermonde matrices are a class of matrices that possess this property but unfortunately are not stable for $n \rightarrow \infty$. This means that the existence of, even low power, additive noise fails the uniqueness. Also, for the cases that our source is approximately sparse, again uniqueness fails. So one must also consider stability towards noise. This is the intuition behind definition of restricted isometry property (RIP) which will be discussed in Chap. 2. Consequently in case of a unique solution, we will end up with the following optimization problem to solve for $\mathbf{s}$:

$$\hat{\mathbf{s}} = \arg\min_{\mathbf{s}} \|\mathbf{s}\|_0 \quad \text{s.t.} \quad \mathbf{y} = A\mathbf{s}, \tag{1.3}$$

where $\|\mathbf{s}\|_0 = |supp(\mathbf{s})|$ denotes the l_0-norm of the vector $\mathbf{s}$.. Unfortunately this problem is NP-Hard and can not be solved for large n. In parallel works [5, 6] it was shown if the sensing matrix satisfies restricted isometry property (RIP) and the signal is sparse enough then (1.2) has a unique solution and more interestingly the solution is viable through solving a convex l_1 optimization problem:

$$\hat{\mathbf{s}} = \arg\min_{\mathbf{s}} \|\mathbf{s}\|_1 \quad \text{s.t.} \quad \mathbf{y} = A\mathbf{s}, \tag{1.4}$$

where $\|\mathbf{s}\|_1 = \sum_i |s_i|$ denotes the l_1-norm of the vector $\mathbf{s}$.

Mathematical proofs for CS theorems is hard to grasp and needs advanced harmonic analysis tools. Here, we try to give some intuitions behind this result. Since l_p norm is not convex for $0 < p < 1$ the problem of (1.3) is considered NP-Hard and the solution is viable only through full search in the feasible region. This is an intractable procedure for large n. The intuition behind replacing (1.3) with (1.4) is to approximate l_0 norm with a convex l_p norm. Obviously the more p is close to zero, the closer the approximated solution is to the real solution. It seems logical to replace l_0 norm with l_1 norm since it is the closest convex norm to l_0 norm. The solution to (1.4) can be found using linear programming algorithms such as basis pursuit (BP) with complexity of $O(n^3)$ [8]. There are various faster methods which will be discussed in Chap. 2.

1.2 Applications of Compressed Sensing

Although theoretical progresses in CS theory is significant, its application has been limited. Very often applications of l_1-norm minimization are considered as applications of CS theory but this is not precisely correct. This area is older and papers such as [9] were published a decade before the derivation of CS theory. Although we can decrease the sampling rate using CS but there are two main barriers which avoid the applicability of CS theory to real world problems. First, one needs to solve the optimization problem (1.4) to reconstruct the signal whereas in ordinary sampling theory, reconstruction is done easily using low-pass filtering. Electronic implementation of (1.4) is more expensive and complex compared to that of a simple low-pass filter. Second, designing the sampling procedure based on CS theory in practice is not simple. Designing a suitable sensing matrix (sampling procedure) is not easy and most known sampling matrices possess stochastic structure. Besides, designing sampling devices where, in each measurement a linear combination of signal samples is measured, is not easy and sometimes is not possible. These limitations restrict the applicability of CS to real world problems. CS can be applied to problems that we really have a problem in sampling issue. Sampling the source is either expensive or limited by the nature or our device. In other words the CS theory is practically useful only in the cases that we are ready to pay the expense of implementing (1.4) to decrease the sampling rate and also designing a measuring device for CS sampling. Here, we review briefly some of those cases.

A promising application of CS is magnetic resonance imaging (MRI) [10]. The MRI sampling device is designed to use magnetic field for imaging human tissues. Due to rules of optics the device measures Fourier coefficients of human body tissues images. On the other hand human tissue images are smooth in time domain and thus sparse in gradient field. Also it is important to decrease the number of samples to decrease the negative effects of magnetic field on human body. If we sample Fourier coefficients partially we have all required conditions, making the use of CS economically logical. This is the reason of applying CS to MRI and several other medical imaging devices. Another direction is to apply CS in wireless network [11], where we have limitations on power consumption and are interested in decreasing the sampling devices (sensors) to save energy. CS also has been applied in radar signal processing, where it is crucial to increase the sampling rate of analog to digital convertors (ADC) [12]. Other applications include error correcting codes [13], biology [14], sparse channel estimation [15], and blind source separation [16]. In this work we will provide two new areas where CS can be applied to, in order to improve the hardware measuring devices.

1.3 Extensions of Compressed Sensing

After the development of the basic CS framework, considerable research has been done on extensions of CS, from defining and proving CS theorems using different mathematical perspectives to providing faster and more efficient CS reconstruction algorithms and investigating related problems such as matrix completion [17]. One possible direction for the extension of compressive sensing is cases where we have additional side information along with source sparsity. It was explained that CS was developed mainly for more efficient signal reconstruction, assuming sparse prior on the source signal. How about cases that we have other types of information about the source signal? for instance we know the source is in the positive orthant ($s \in \mathbb{R}^n_+$) or we have some information about the source structure. Can we use the side information and combine it with the sparsity prior to further decrease the sampling rate and improve signal reconstruction? Answering this question has resulted in several directions for extending CS. Depending on the type of side information, several extensions on CS theory has been reported in the literature.

One type of side information is information about the source structure. For instance along with sparsity the non-zero elements may have a pattern. In [18], the authors have provided CS framework for block-sparse signals, i.e., sparse signals in which non-zero element appear in blocks rather than individually. Block sparse signals are a suitable model for pulse-shaped signals such as radar signals. Authors have derived corresponding adopted CS theorems and recovery algorithms for this case. Some applied signals such as image/video signals have spatial/time structure. Normally in a natural image the value of a pixel has correlation with neighboring pixels. Apart from spatial correlation, values of a pixel in a video stream are correlated over time. In [19] the authors have tried to exploit the frame correlation to improve signal recovery of images/video signals. In the case of an image, the image is split into blocks of fixed size, which are all sparse in the same domain, and then each block is recovered using information extracted from neighboring blocks. In case of a video stream, a previous frame is used to help recovery of consecutive frames.

Another type of information is about the source probability distribution. Some signals such as texture images can be modeled better via probabilistic models rather than deterministic models. In [20], the authors provide a CS recovery algorithm when prior information about the probability of each nonzero entry of the source is in hand. Others have tried to adopt probabilistic graphical model message passing algorithms such as belief propagation to result in faster CS recovery algorithms when the source distribution is known [21, 22]. First an equivalent graphical model to a CS problem, inspired from error correction codes, has been developed and next, message passing has been used to solve for the source. This approach has provided some fast recovery algorithms compared to CS general reconstruction algorithms. Research is going on to develop non-parametric recovery algorithms using this approach in order to extend it to the general case [22].

The third type of information is knowledge about the feasible region. Imagine we have information which limits the feasible region of (1.3), which is automorphic to null(A). For instance $c \in \mathbb{R}^n_+$. The previous information types dealt with the nature of the source, so the uniqueness of the solution still holds for (1.3), whereas when the feasible region changes uniqueness of the solution may fail. Consequently one must first check the uniqueness of the solution and then provide the corresponding sparse recovery algorithm. In [23] authors have provided an algorithm for non-negative sources. Uniqueness of the solution is proved and one of the common CS recovery algorithms has been altered to work for such sources more efficiently. A more general case happens when another convex constraint is added to (1.3), i.e., $Bc = b$, $B \in \mathbb{R}^{m' \times n}$, $b \in \mathbb{R}^{m'}$ (note inequality conditions like $s > 0$ can be transformed to equality condition). In the current work, we focus on such a case.

Special case of this problem has been studied in the literature. In [24] derivative compressive sampling is introduced. It is assumed that the source signal is a gradient field. Next, it has been shown the problem can be transformed to an equivalent CS problem and then solved using a general sparse recovery algorithm. In the current work, uniqueness of the solution for the more general case is studied. Then it will be shown for this case equivalence of l_0-norm and l_1-norm minimization solutions still holds. Furthermore, a more efficient algorithm is provided for solution.

1.4 Organization

This chapter covered a brief review of the compressed sensing (CS) theory. Chapter 2 covers a more detailed survey on CS theory and its mathematical foundations. The classic framework of CS is provided in Sects. 2.1 and 2.2. Section 2.3 provides a review on a more recent mathematical foundation for CS based on spherical section property. This framework is easier to grasp and can be adopted for our problem, as described in Chap. 3. Section 2.4 reviews CS reconstruction methods. In Chap. 3, the problem of CS in the presence of side information about feasible region is studied. After formulating the problem, uniqueness and stability of the l_1-norm reconstruction are provided in Sect. 3.2. The next three chapters are devoted to applications of the developed scheme. In Chap. 4 the problem of deburring in optical imaging is studied. It is shown how the provided scheme can be used to improve performance of interferometer devices. In Chap. 5 the problem of surface reconstruction in the gradient field is studied and Chap. 6 addresses the problem of diffusive field reconstruction. Numerical simulations for these applications confirm the effectiveness and usefulness of the proposed method.

References

1. C.E. Shannon, Communication in the presence of noise. Proceedings of the IRE **37**(1), 10–21 (1949)
2. A.J. Coulson, A generalization of nonuniform bandpass sampling. IEEE Transactions on Signal Processing **43**(3), 694–704 (1995)
3. K.L. Clarkson, P.W. Shore, Applications of random sampling in computational geometry. Springer, Discrete and Computational Geometry **4**(1), 387–421 (1989)
4. R.O. Schmidt, Multiple emitter location and signal parameter selection. IEEE Transactions on Antenna and Propagation **34**, 276–280 (1986)
5. E.J. Candés, J. Romberg, T. Tao, Robust uncertainty principles: exact signal reconstruction from highly incomplete frequency information. IEEE Transactions on Information Theory **52**(2), 489–509 (2006)
6. Y. Tsaig, D.L. Donoho, Compressed sensing. IEEE Transactions on Information Theory **52**, 1289–1306 (2006)
7. R. G. Baraniuk. Compressive sensing. In Proceedings of 42nd Annual Conference on Information Sciences and Systems, Princeton, NJ, USA, March 2008

8. J. Tropp, Recovery of short linear combinations via l_1 minimization. IEEE Transactions on Information Theory **90**(12), 4655–4666 (2005)
9. R. Tibshirani, Regression shrinkage and selection via the Lasso. Journal of the Royal Statistical Society, Series B **58**, 267–288 (1994)
10. M. Lustig, D.L. Donoho, J.M. Santos, J.M. Pauly, Compressed sensing MRI. IEEE Signal Processing Magezine **25**(2), 72–82 (2008)
11. W. Bajwa, J. Haupt, A. Sayeed, and R. Nowak. Compressive wireless sensing. In Proceedings of the 5th International Conference on Information Processing in Sensor Networks, IPSN '06, pages 134–142, 2006
12. M. Herman, T. Strohmer, High-resolution radar via compressed sensing. IEEE Transactions on Signal Processing **57**(6), 2275–2284 (June 2009)
13. R. Chartrand. Nonconvex compressed sensing and error correction. In Proceedings of the 2007 IEEE International Conference on Acoustics, Speech and, Signal Processing, pp. 889–892, 2007
14. M.A. Sheikh, O. Milenkovic, S. Sarvotham, R (Compressed sensing DNA microarrays, G. Baraniuk, 2007)
15. C.R. Berger, Z. Wang, J. Huang, S. Zhou, Application of compressive sensing to sparse channel estimation. IEEE Communications Magezine **48**(11), 164–174 (2010)
16. T. Blumensath, M. Davies, *Compressed sensing and source separation* (In International Conference on Independent Component Analysis and Blind, Source Separation, 2007)
17. E.J. Candès, B. Recht, Exact matrix completion via convex optimization. Foundations of Computational Mathematics **9**(6), 717–772 (2009)
18. Y. Eldar, P. Kuppinger, H. Bölcskei, Block-sparse signals: Uncertainty relations and efficient recovery. IEEE Transactions on Signal Processing **58**(6), 3042–3054 (2010)
19. V. Stanković, L. Stanković, and S. Cheng. Compressive image sampling with side information. In Proceedings of the 16th IEEE International Conference on Image Processing, ICIP'09, pages 3001–3004, 2009
20. M.A. Khajehnejad, W. Xu,, A.S. Avestimehr, and B. Hassibi. Weighted l_1 minimization for sparse recovery with prior information. In Proceedings of the 2009 IEEE international Symposium on Information Theory, ISIT'09, pages 483–487, Piscataway, NJ, USA, 2009
21. D. Baron, S. Sarvotham, R.G. Baraniuk, Bayesian compressive sensing via belief propagation. IEEE Transactions on Signal Processing **58**(1), 269–280 (2010)
22. D.L. Donoho, A. Maleki, A. Montanari, Message-passing algorithms for compressed sensing. Proceedings of National Academy of Science **106**(45), 18914–18919 (2009)
23. M.A. Khajehnejad, A.G. Dimakis, W. Xu, B. Hassibi, Sparse recovery of nonnegative signals with minimal expansion. IEEE Transactions on Signal Processing **59**(1), 196–208 (2011)
24. M. Hosseini, O. Michailovich, *Derivative compressive sampling with application to phase unwrapping* (In Proceedings of EUSIPCO, Glasgow, UK, August, 2009)

Chapter 2
Compressed Sensing

In this chapter compressed sensing is introduced in more details. Gelfand's width, which is a pure mathematical concept with close connection to CS, is introduced in Sect. 2.1. CS in restricted isometry perspective is considered in Sect. 2.2. Section 2.3 covers a review on spherical section property, which will be used in the next Chapter.

2.1 Gelfand's Width

Some mathematical ideas that are used in CS originally came from the Harmonic Analysis literature. In this section we introduce Gelfand's width and show how it is connected with CS theory. Let $S \subset \mathbb{R}^n$ and $m < n \in \mathbb{N}$. Assume $\mathbb{R}^n$ is equipped with l_p-norm.

Definition 2 *Gelfand's width for this set is defined as:*

$$d^m(S)_p = \inf_K \sup\{\|x\|_p | x \in S \cap K\} = \inf_K \sup_{x \in S \cap K} \|x\|_p, \ p \geq 1, \qquad (2.1)$$

where infimum is taken over all $n - m$ dimensional subspace K of $\mathbb{R}^n$. Assume S be bounded such that:

$$\forall s \in S : -s \in S \qquad (2.2)$$
$$\exists a \in \mathbb{R}^n : S + S \subset aS$$

For instance if $S = \{x \in \mathbb{R}^n | \|x\| < 1\}$, then assuming $a = 2$, this set satisfies (2.2). Now assume we sample elements of S with a sampling matrix $\Phi \in \mathbb{R}^{m \times n}$. Also let D be an operator (possibly nonlinear) which is used for reconstructions:

$$\mathbf{y} = \Phi\mathbf{x}$$
$$\hat{\mathbf{x}} = D(\mathbf{y}) \qquad (2.3)$$

M. Rostami, *Compressed Sensing with Side Information on the Feasible Region*,
SpringerBriefs in Electrical and Computer Engineering,
DOI: 10.1007/978-3-319-00366-5_2, © The Author(s) 2013

The error reconstruction in this sampling/reconstruction system over the set S would be:

$$E(S, \Phi, D) = \sup_{\mathbf{x} \in S} \|\mathbf{x} - \hat{\mathbf{x}}\|_p = \sup_{\mathbf{x} \in S} \|\mathbf{x} - D(\Phi \mathbf{x})\|_p \qquad (2.4)$$

We are interested in finding a (Φ, D) pair such that $E(S, \Phi, D)$ is minimized. The best possible performance in this framework is given by:

$$E(S) = \inf_{\Phi, D} E(S, \Phi, D). \qquad (2.5)$$

As we know $\dim(\text{Null}(\Phi)) = n - m$ and thus $\text{null}(\Phi)$ is a $n - m$ dimensional subspace of $\mathbb{R}^n$ and can be considered an instance of K in the definition of Gelfand's width of S:

$$d^m(S)_p = \inf_K \sup_{\mathbf{x} \in S \cap K} \|\mathbf{x}\|_p \leq \sup_{\mathbf{x} \in S \cap \text{null}(\Phi)} \|\mathbf{x}\|_p. \qquad (2.6)$$

On the other hand:

$$\forall \mathbf{x} \in S \cap \text{null}(\Phi) : D(\mathbf{y}) = D(\Phi \mathbf{x}) = D(0) = D(-\Phi \mathbf{x}) \qquad (2.7)$$

Now note:

$$\|\mathbf{x} - D(\Phi \mathbf{x})\|_p + \| - \mathbf{x} - D(-\Phi \mathbf{x})\|_p \geq \|\mathbf{x} - D(0) - \mathbf{x} + D(0)\|_p = 2\|\mathbf{x}\|_p \rightarrow \qquad (2.8)$$

$$\|\mathbf{x} - D(0)\|_p \geq \|\mathbf{x}\|_p \quad \text{or} \quad \| - \mathbf{x} - D(0)\|_p \geq \|\mathbf{x}\|_p.$$

Thus for any $\mathbf{x} \in S \cap \text{null}(\Phi)$, there exists an element $\mathbf{x}' \in S \cap \text{null}(\Phi)$ such that: $\|\mathbf{x}' - D(\Phi \mathbf{x}')\|_p \geq \|\mathbf{x}\|_p$. Consider this fact:

$$E(S, \Phi, D) \geq \sup_{\mathbf{x} \in S \cap K} \|\mathbf{x} - D(\Phi \mathbf{x})\|_p \geq \sup_{\mathbf{x} \in S \cap \text{null}(\Phi)} \|\mathbf{x} - D(\Phi \mathbf{x})\|_p \geq \sup_{\mathbf{x} \in S \cap \text{null}(\Phi)} \|\mathbf{x}\|_p. \qquad (2.9)$$

From (2.6) and taking infimum on (2.9) one can conclude:

$$E(S)_p \geq d^m(S)_p. \qquad (2.10)$$

Now assume $K \subset \mathbb{R}^n$ with $\dim(K) = n - m$. Let $\{\mathbf{v}_1, \ldots, \mathbf{v}_m\}$ be a basis for orthogonal complement of K ($K^\perp$). Form the sampling matrix $\Phi = [\mathbf{v}_1, \ldots, \mathbf{v}_m]^T$. Also we define the reconstruction operator D as follows:

$$D(\mathbf{u}) = \begin{cases} \mathbf{a} & \text{if } \mathbf{u} \in \Phi S \\ \mathbf{b} & \text{if } \mathbf{u} \notin \Phi S \end{cases} \qquad (2.11)$$

where $\mathbf{a} \in S$ is arbitrary such that $\mathbf{u} = \Phi \mathbf{a}$ and $\mathbf{b}$ is a randomly chosen vector in S. With these assumptions on (Φ, D) we calculate $E(S)_p$. Let $\mathbf{x} \in S$:

$$\Phi(\mathbf{x} - D(\Phi\mathbf{x})) = \Phi\mathbf{x} - \Phi D(\Phi\mathbf{x}) = \Phi\mathbf{x} - \Phi\mathbf{x} = 0, \tag{2.12}$$

which yields $\mathbf{x} - D(\Phi\mathbf{x}) \in \text{null}(\Phi) \equiv K$. Also from (2.2):

$$\exists a \in \mathbb{R} : \frac{\mathbf{x} - D(\Phi\mathbf{x})}{a} \in S \to \frac{\mathbf{x} - D(\Phi\mathbf{x})}{a} \in S \cap K. \tag{2.13}$$

Consequently:

$$E(S, \Phi, D)_p = a \sup_{\mathbf{x} \in S} \left\| \frac{\mathbf{x} - D(\Phi\mathbf{x})}{a} \right\|_p \le a \sup_{\mathbf{x} \in S \cap K} \|\mathbf{x}\|_p \to$$

$$\inf_{\Phi, D} E(S, \Phi, D)_p \le a \inf_{K} \sup_{\mathbf{x} \in S \cap K} \|\mathbf{x}\|_p \to E(S)_p \le a d^m(S)_p. \tag{2.14}$$

Overall from (2.10) and (2.14):

$$d^m(S)_p \le E(S)_p \le a d^m(S)_p. \tag{2.15}$$

This is an important result and shows how the reconstruction error over the set S is related to Gelfand's width of S. In other words, the best reconstruction performance in CS is bounded by Gelfand's width. Unfortunately finding Gelfand's width of a set in the general case is an open problem and only for special instances of S, such as unit ball, Solutions have been found. Advances in this area provide a strong mathematical imbed for CS theory.

The central question is what Φ, D pair would satisfy the bounds given by Gelfand's width? Independently, in [1, 2] sufficient condition on the sensing matrix was provided. The authors introduced the concept of restricted isometry property (RIP) and used this concept to provide theorems for unique and stable source reconstruction and prove the CS theorems. They showed that the random sensing matrices with i.i.d. Gaussian or Bernoulli entries satisfy the required conditions and efficient decoding D, can be accomplished by linear programming as in (1.4) (this reconstruction method has been provided before through empirical approach).

2.2 Restricted Isometry Property and Coherence

The classical theory of CS [1, 2] uses the concept of RIP. As discussed in Chap. 1, if the source is k-sparse, then if any combination of $2k$ columns of A is linearly independent, then the solution of (1.2) would be unique. Having this in mind, the restricted isometry property (RIP) is defined as follows:

Definition 3 *Restricted Isometry Property*
We say an arbitrary matrix A, satisfies RIP of order k with constant $0 \le \delta_k < 1$, if for all k-sparse vectors $\mathbf{x}$:

$$1 - \delta_k \leq \frac{\|Ax\|_2^2}{\|x\|_2^2} \leq 1 + \delta_k. \tag{2.16}$$

This means that k-sparse sources not only will not lay in the null-space of A, but also will have a distance δ_k with this space. This condition is stronger compared to linear independency of any $2k$ columns of A and in return is also stable towards noise. In other words it means that all sub-matrices of A with at most k columns are well-conditioned. The constant $0 \leq \delta_k < 1$ measures closeness of the sensing operator to an orthonormal system. From discussions in Chap. 1, one concludes if A satisfies RIP with $0 \leq \delta_{2k} < 1$ then the solution of (1.2) is unique and can be recovered through solving (1.3). But for practical applications equivalence of the solutions of (1.3) and (1.4) is essential.

Historically CS results are developed using RIP and over the time the conditions and bounds on theorems are improved. The following two theorems are two main state-of-the-art results based on the RIP approach [3].

Theorem 2.2.1 (Noiseless Recovery) *Consider the system* (1.2) *with the unique solution* s, *assume* $\delta_{2k} < \sqrt{2} - 1$. *Let* $\hat{s}$ *be the solution to* (1.4), *then:*

$$\|s - \hat{s}\|_1 \leq C_0 \|s - s_k\|_1$$

and

$$\|s - \hat{s}\|_2 \leq C_0 \frac{1}{\sqrt{k}} \|s - s_k\|_1$$

where s_k *is a* k-*sparse approximation of* s *and* C_0 *is a global constant.*

Note that for the case the source is exactly k-sparse, this theorem states the recovery is exact. The next theorem states the condition for robustness towards noise.

Theorem 2.2.2 (Noiseless Recovery) *Consider the system* $y = As + n$ *such that* $\|n\|_2 < \epsilon$, *assume* $\delta_{2k} < \sqrt{2} - 1$. *Let* $\hat{s}$ *be the solution to* (1.4), *then:*

$$\|s - \hat{s}\|_2 \leq C_0 \frac{1}{\sqrt{k}} \|s - s_k\|_1 + C_1 \epsilon$$

with the same constant C_0 *as in the previous theorem and another global constant* C_1.

Proofs of these theorems are complicated and based on advanced real analysis mathematics. Interested readers may refer to [1–3] for details.

RIP condition on the sensing matrix is a standard approach in CS theory, but unfortunately its practical benefits is limited. Calculating RIP for a general matrix is a NP-hard problem and only has been done for special cases. Using random matrix theory, existence of such matrices have been proven for $m > O(k \log(\frac{n}{k}))$

for any desired $\delta_k \in (0, 1)$ but even in such cases building such matrices is an independent issue. Note these theorems require RIP condition but we will discuss that RIP condition is only a sufficient condition and is not a necessary condition for accurate l_1-recovery. On the other hand, it is also not a complete concept to study CS.

An important quantity in designing the sensing matrix is mutual coherence.

Definition 4 *Mutual Coherence*
  Let $A \in \mathbb{R}^{n \times m}$, the mutual coherence μ_A is defined by:

$$\mu_A = \max_{i \neq j} \frac{|\langle a_i, a_j \rangle|}{\|a_i\| \|a_j\|}$$

where a_i, a_j denote two distinct columns of A.

A small coherence implies of closeness of the sensing matrix to a normal matrix. If a matrix possesses a small mutual coherence, then it also satisfies the RIP condition. It means that coherence is a stronger condition. On the other hand the complexity of calculating the coherence is $O(n^2)$ and thus is tractable. According to Welch inequality [4]:

$$\mu_A \geq \sqrt{\frac{n}{m(n-m)}} \tag{2.17}$$

This implies for $n \gg m$, $\mu_A \geq \frac{1}{\sqrt{m}}$. Consequently if we want to design sensing matrices which satisfy RIP condition using mutual coherence, then $m > O(k^2)$ which is much greater than $m = O(k \log(\frac{n}{k}))$ bound for which existence of proper sensing matrices has been proven. But due to computational complexity issues, it is the only proper tool for this purpose.

Next section covers the new paradigm for compressive sensing [5–7]. This approach uses a completely different approach based on studying the nullspace of the sensing matrix using spherical section property.

2.3 Spherical Section Property

Analysis of compressive sensing based on RIP requires advanced mathematical tools, but this approach is not necessary to develop compressive sensing [5, 6]. Moreover, it is not a required condition for exact recovery.

Consider the problem of (1.2). The pair (A, y) carries the information in CS framework. Consider an invertible matrix, B. It is trivial that the system $BAs = By$ is equivalent to the system $As = y$. Thus the pair (BA, By) carries the same information as (A, y). But the RIP of A and BA can be vastly different. For any CS problem one can choose B to make RIP of BA significantly bad, regardless of RIP of A [6]. RIP is a strong condition on sensing matrix and practical and experimental results confirm it is not a necessary condition for main theorems of CS to hold. This motivates the derivation of CS in a more simple and general approach based on spherical section property (SSP) [5, 6]. Interestingly this approach is simpler and some of the main results of CS theorems in RIP context can be derived easier using spherical section property. Here we briefly describe CS theory in this context and follow the approach of [6] in proving the main theorems.

Definition 5 *Spherical Section Property (SSP) Let* $m, n \in \mathbb{N}$ *such that* $n > m$ *and* V *be an* $n - m$ *dimensional subspace of* $\mathbb{R}^n$. *This subspaces is said to have spherical section property with constant* Δ, *if* $\forall s \in V$:

$$\frac{\|s\|_1}{\|s\|_2} \geq \sqrt{\frac{m}{\Delta}}$$

Here, Δ *is called the distortion of* V.

Note if we consider the nullspace of a sensing matrix as the subspace in this definition, for an invertible matrix $\Delta = 0$. Similar to RIP approach the following theorems are developed.

Theorem 2.3.1 (Noiseless Recovery) *Suppose* null(A) *has the* Δ-*spherical section property. Let* $\hat{s}$ *be a nonzero vector such that:* $A\hat{s} = y$.

1. *Provided that:* $\|\hat{s}\|_0 \leq \frac{m}{3\Delta}$, $\hat{s}$ *is the unique vector satisfying* $As = y$ *and* $\|s\|_0 \leq \frac{m}{3\Delta}$.
2. *Provided that:* $\|\hat{s}\|_0 \leq \frac{m}{2\Delta} \leq \frac{n}{2}$, $\hat{s}$ *is the unique solution to the optimization problem* (1.4).

Proof 1

1. *First define the vector* sign$(s) = [\text{sign}(s_i)]$. *According to the Cauchy-Schwarz inequality:*

$$|\langle \text{sign}(s), s \rangle| \leq \|\text{sign}(s)\|_2 \|s\|_2 \rightarrow \sum_i |s_i| \leq \sqrt{|supp(s)|}\|s\|_2 \rightarrow \|s\|_1$$

$$\leq \sqrt{\|s\|_0}\|s\|_2 \tag{2.18}$$

Now assume v *be a second solution which is more sparse compared to* $\hat{s}$ *and* $\|v\|_0 = m_1$. *Let* $w = v - \hat{s}$. *Note,* $w \neq 0$ *and* $w \in \text{Null}(A)$, *then:*

$$\|w\|_0 \leq \|v\|_0 + \|\hat{s}\|_0 \leq m_1 + \frac{m}{3\Delta} \xrightarrow{(2.18)} \frac{\|w\|_1}{\|w\|_2} \leq \sqrt{m_1 + \frac{m}{3\Delta}} \tag{2.19}$$

$$\sqrt{\frac{m}{\Delta}} \leq \sqrt{m_1 + \frac{m}{3\Delta}} \rightarrow \frac{2m}{3\Delta} \leq m_1,$$

this a contradiction and shows v *is not sparse enough and uniqueness of the solution results.*

2. *Again assume* v *be a second solution to (1.4) such that* $\|v\|_1 \leq \|\hat{s}\|_1$ *and let* $w = v - \hat{s}$, $S = supp(\hat{s})$, $\bar{S} = \{1, ..., n\} - S$, *and* w_S *to be the projection of* w *on* S:

$$\|v\|_1 = \|w + \hat{s}\|_1 = \|w_S + \hat{s}_S\|_1 + \|w_{\bar{S}} + \hat{s}_{\bar{S}}\|_1 = \|w_S + \hat{s}_S\|_1 + \|w_{\bar{S}}\|_1 \geq \tag{2.20}$$

$$\|\hat{s}_S\|_1 - \|w_S\|_1 + \|w_{\bar{S}}\|_1 = \|\hat{s}\|_1 - \|w_S\|_1 + \|w_{\bar{S}}\|_1,$$

now since $\|v\|_1 \leq \|\hat{s}\|_1$, *one concludes* $\|w_{\bar{S}}\|_1 \leq \|w_S\|_1$.

Note $w \in$ null(A), *now we want to calculate maximum value of the ratio* $\frac{\|w\|_1}{\|w\|_2}$. *This problem is invariant under scaling of* w, *thus we set* $\|w\|_2 = 1$ *and also we can assume* w *lays in the positive orthant (since the element signs would not change the norm value). We will have the following optimization problem:*

$$\max \quad w_1 + \cdots + w_n$$
$$s.t.: \quad 0 \leq w_i, \tag{2.21}$$
$$\sum_{i \in \bar{S}} w_i \leq \sum_{i \in S} w_i$$

The second constraint comes from the inequality we derived before. This problem is a convex optimization instance, so we can exhibit the maximizer in closed form if we can exhibit the solution to the KKT condition [8]. Let

$$w_i = \begin{cases} a = \dfrac{\sqrt{\|\hat{s}\|_0(n - \|\hat{s}\|_0)/n}}{\|\hat{s}\|_0}, & i \in S \\[3mm] b = \dfrac{\sqrt{\|\hat{s}\|_0(n - \|\hat{s}\|_0)/n}}{\|n - \hat{s}\|_0}, & i \in \bar{S} \end{cases} \tag{2.22}$$

It is easy to check that this point lays in the feasible region. The KKT multipliers are the solutions to the system:

$$\begin{cases} \lambda_1 + 2\lambda_2 b = 1 \\ -\lambda_1 + 2\lambda_2 a = 1 \end{cases} \rightarrow \begin{cases} \lambda_1 = \dfrac{a - b}{a + b} \\[3mm] \lambda_2 = \dfrac{1}{a + b} \end{cases} \tag{2.23}$$

So both multipliers are positive if $\|\hat{s}\|_0 \leq \|n - \hat{s}\|_0$. *Thus the objective value of (2.21) would be* $\sqrt{\dfrac{\|\hat{s}\|_0(n - \|\hat{s}\|_0)}{n}}$ *and consequently* $\dfrac{\|w\|_1}{\|w\|_2} \leq \sqrt{\dfrac{\|\hat{s}\|_0(n - \|\hat{s}\|_0)}{n}}$. *On the other hand* $w \in$ null(A), *which concludes:*

$$\sqrt{\frac{m}{\Delta}} \le \sqrt{\frac{\|\hat{s}\|_0 (n - \|\hat{s}\|_0)}{n}} \le \sqrt{\|\hat{s}\|_0} \rightarrow \frac{m}{\Delta} \le \|\hat{s}\|_0, \tag{2.24}$$

which contradicts the assumption and results the proof.

The second theorem considers stability towards noise.

Theorem 2.3.2 Noisy Recovery *Suppose* null(A) *has the* Δ-*spherical section property. Let* $\hat{s}$ *be the minimizer of* (1.4). *Then for every* $\bar{s} \in \mathbb{R}^n$ *and* $\forall k < \min(\frac{m}{16\Delta}, \frac{n}{4})$:

$$\|\hat{s} - \bar{s}\|_1 \le 4\|\bar{s}_k - \bar{s}\|_1, \tag{2.25}$$

where s_k *denotes the k-sparse approximation of* s.

Proof 2 *Let* $w = \hat{s}_k - \bar{s}$, *so* $w \in$ null(A):

$$\begin{aligned}
\|\hat{s}\|_1 &= \|\bar{s} + w\|_1 = \\
&\|\bar{s}_S + w_S\|_1 + \|\bar{s}_{\bar{S}} + w_{\bar{S}}\|_1 \ge \\
&\|\bar{s}_S\|_1 - \|w_S\|_1 - \|\bar{s}_{\bar{S}}\|_1 + \|w_{\bar{S}}\|_1 \ge \\
&\|\bar{s}\|_1 - \|w_S\|_1 + \|w_{\bar{S}}\|_1 - 2\|\bar{s}_{\bar{S}}\|_1,
\end{aligned} \tag{2.26}$$

Since $\hat{s}$ *is the minimizer of* (1.4) *we conclude:*

$$\|w_{\bar{S}}\|_1 \le \|w_S\|_1 + 2\|\bar{s}_{\bar{S}}\|_1. \tag{2.27}$$

Now define: $R = \frac{\|w\|_1}{\|\bar{s} - s_k\|_1}$. *To obtain the result, it is enough to find an upper bound for R* ($R \le 4$). *We substitute R in* (2.27):

$$\|w_{\bar{S}}\|_1 \le \|w_S\|_1 + 2\|w\|_1/R \rightarrow \|w_{\bar{S}}\|_1 \le \|w_S\|_1 + 2(\|w_S\|_1 + \|w_{\bar{S}}\|_1)/R \rightarrow \tag{2.28}$$

$$(1 - 2/R)\|w_{\bar{S}}\|_1 \le (1 + 2/R)\|w_S\|_1.$$

Now note if $1 - 2/R \ge 0$, *then* $R \le 2 \le 4$ *and the proof results, so let* $1 - 2/R > 0$. *Then from* (2.28): $\|w_{\bar{S}}\|_1 \le \frac{1+2/R}{1-2/R}\|w_S\|_1$. *Assuming* $\gamma = \frac{1+2/R}{1-2/R}$ ($\gamma \le 3$) *and in exactly the same approach as in the previous theorem one can conclude (for details refer to* [6]):

$$\frac{\|w\|_1}{\|w\|_2} \le \gamma + \gamma\sqrt{\frac{k(n-k)}{k + 9(n-k)}} \xrightarrow{w \in \text{null}(A)} \sqrt{\frac{m}{\Delta}} \le \gamma + \gamma\sqrt{\frac{k(n-k)}{k + 9(n-k)}} \tag{2.29}$$

$$\xrightarrow{n-k \le ((9(n-k)+k)/9)} \sqrt{\frac{m}{\Delta}} \le (\gamma + 1)\sqrt{k} \xrightarrow{k \le \frac{m}{16\Delta}} 3 \ge \gamma.$$

On the other hand the assumption was $\gamma \leq 3$ and thus $\gamma = 3$. Consequently $R = 2$ which results in the desired bound on R and the result follows.

These two theorems establish CS theory but in SSP context and similarly state uniqueness and stability of l_1-norm solution for a CS problem. The results are derived in a much simpler approach compared to RIP context [1, 2]. It is interesting to note that the main results which are derived in RIP approach can be rederived in SSP context. For instance the error bound in Theorem 2.3.2 has been derived in RIP context, too. Also, it has been shown the Gaussian random matrices have spherical section property and are good choice for sensing matrix [5]. Furthermore, as it will be discussed this approach is a better embed for considering cases when we have side information on the feasible region.

2.4 Reconstruction Methods

In this section a brief review on CS reconstruction methods is given. Nowadays one of the limitation of using CS is the low-speed of the reconstruction methods with high dimensional data. Improving the performance of reconstruction methods is an active research area.

2.4.1 Minimization of l_1-norm

Historically l_1-norm minimization is the main approach for CS reconstruction algorithms. Main CS theorems state robustness of the l_1-norm minimization towards additive noise and also system noise. The importance of l_1-norm is that, it is a continuous convex function, so convex optimization tools can be applied to the problem. The more important fact is that l_1-norm minimization problem can be formulated as a linear programming problem. Let $A' = [A, -A]$, $s' = [s_1; s_2]$, $s = s_1 - s_2$:

$$\min[1; 1]^T s' \quad \text{s.t.} A's' = y0, s' \geq 0, \tag{2.30}$$

where 1 is an all-ones column vector and $(\cdot)^T$ denotes matrix transposition. Consequently well-known linear programming algorithms such as Simplex and Interior Point methods can be used with complexity of $O(n^3)$. One group of successful algorithms in this class is Basis Pursuit [9].

Although linear programming methods can find the solution in finite time but for many practical applications $O(n^3)$ is not a tractable time. Specially in image processing applications in which $n = O(10^5)$ for a typical image.

2.4.1.1 Thresholding Algorithms

Some iterative methods have been introduced to decrease the computational complexity of l_1-norm minimization. In these methods an iterative sequence of vectors is produced, which converges to the solution through iterations. Although convergence to the exact solution is more time consuming compared to linear programming methods, these methods quickly converge to a very good approximate of the solution.

It can be shown that for a proper selection of λ, the optimization problem (1.4) is equivalent to the following unconstrained problem:

$$\hat{\mathbf{s}} = \arg\min_{\mathbf{s}} \frac{1}{2}\|\mathbf{y} - A\mathbf{s}\|_2^2 + \lambda\|\mathbf{s}\|_1 \tag{2.31}$$

Since this problem is unconstrained one can use steepest descend or conjugate gradient approaches to derive an iterative relation. Although l_1-norm is not a smooth function but concept of subderivative enables us to apply a similar procedure to steepest descend on (2.31) (more discussions is given in Chap. 3). Upon choosing a proper initial value, the iterative relation will converge to the minimizer of (1.4). Several algorithms have been developed for this purpose [10, 11]. In the current note we work with image signals and thus we have used one of the-sate-of-the-art iterative methods for reconstruction [12, 13].

The iterative formula for iterative hard thresholding (IHT) algorithm is as follows:

$$\mathbf{s}^{i+1} = \mathcal{G}(\mathbf{s}^i - A^T(A\mathbf{s}^i - \mathbf{y})), \tag{2.32}$$

where $\mathcal{G}(\cdot)$ is a thresholding function:

$$\mathcal{G}(x) = \begin{cases} 0 & |s_i| \leq \sqrt{\lambda} \\ s_i & |s_i| \geq \sqrt{\lambda} \end{cases}, \tag{2.33}$$

The main advantage is that each iteration only involves multiplication of vectors and A and A^T, followed by thresholding. So the sensing matrix can be defined only as an operator and it is not even required to store the sensing matrix. This is much simpler than linear programming. Note the threshold in this algorithm is constant in all iterations. A class of successful methods is the iterative shrinkage thresholding algorithms (ISTA) which improve IHT through using an adaptive thresholding function. The iterative step is as follows:

$$\mathbf{s}^{i+1} = \mathcal{H}_{\lambda\delta}(\mathbf{s}^i - \delta A^T(A\mathbf{s}^i - \mathbf{y})), \tag{2.34}$$

where δ is a parameter for step size and $\mathcal{H}(\cdot)$ is a soft shrinkage threshold function:

$$\mathcal{H}_\lambda(s_i) = (|s_i| - \lambda)_+ \text{sign}(x_i). \tag{2.35}$$

FISTA algorithm [12] further improves ISTA by involving the solutions of the two previous iterations in each step.

2.4.2 Greedy Algorithms

Greedy algorithms generally solve a problem in a number of steps (in CS problem, mainly the number of steps is equal to the sparsity level k). In each step the best selection (in CS problem, normally the best column of the sensing matrix) is done without considering the future steps. Consequently the result is not always the real solution but this approach provides acceptable results in compressive sensing reconstruction.

A simple algorithm of this class is Matching Pursuit. An equivalent representation for compressive sensing is:

$$\mathbf{y} = \sum_{i=1}^{n} \mathbf{a}_i s_i, \qquad (2.36)$$

where $\mathbf{a}_i$ is the ith column of A. If we have a k-sparse source, CS in this context can be interpreted as finding the k related columns of A and corresponding s_i's. Matching Pursuit approximates the source in k step. In each step one column of A is revealed and then the corresponding s_i is revealed by solving a least square problem. In the first step the inner product of $\mathbf{y}$ and all $\mathbf{a}_i$'s are calculated ($\langle \mathbf{y}, \mathbf{a}_i \rangle$). Then the column $\mathbf{a}_j$ with the maximum absolute value of $\langle \mathbf{y}, \mathbf{a}_i \rangle$ is selected as an active column in (2.36) and $s_i = \langle \mathbf{y}, \mathbf{a}_j \rangle$. Thus the first term in (2.36) is known. Let this approximate of $\mathbf{s}$ be $\mathbf{s}^{(1)}$. The next steps are done similarly, only in each step we update the value of $\mathbf{y}$ as follows:

$$\mathbf{y}^{(i+1)} = \mathbf{y}^{(i)} - s_i \mathbf{a}_j. \qquad (2.37)$$

The main disadvantage in this approach is that it is assumed that columns of A are orthogonal which is not the case for most sensing matrices. Orthogonal Matching Pursuit (OMP) [14] improves this method via updating the found s_i's in each step. Since this approach uses similarity of $\mathbf{a}_i$'s and the residual vector of (2.37), mutual coherency of the sensing matrix plays an important role. Faster algorithms such as Compressive Sampling Matched Pursuit (CoSaMP) [15] improves the algorithm via a look on future steps. Overall, this class of reconstruction methods are fast but do not necessarily find the real solution.

2.4.3 Norm Approximation

This class approximate l_0-norm via a differentiable function and then use methods such as steepest descend for minimization. For instance smoothed l_0 (SL0) algorithm [16] approximates the l_0-norm as follows:

$$\|\mathbf{s}\|_0 \approx g(\mathbf{s}) = n - \sum_{i=0}^{n} f_\sigma(s_i), \tag{2.38}$$

where $f_\sigma(\cdot)$ is defined as:

$$f_\sigma(s) = e^{-\frac{s^2}{2\sigma^2}}, \tag{2.39}$$

and $\sigma \in \mathbb{R}^+$ is a small constant. The parameter σ determines the closeness to the l_1-norm and smoothness of the approximation, as $\sigma \to 0$ then $g(\mathbf{s}) \to \|\mathbf{s}\|_0$. The function $g(\cdot)$ is continuous and differentiable and thus steepest descend methods can be applied directly to find the minimizer of $g(\cdot)$. For a proper selection of σ, it may be possible to find the global minimizer of (1.3). Experiments have shown that this method is faster than l_1-norm minimization methods but again for large scale systems it is not applicable.

2.4.4 Message Passing Reconstruction Algorithms

Graphical Models is an active research area with a wide range of applications. Recently fast iterative methods based on graphical models have been used in convex optimization problems [17, 18]. The connection between belief propagation (BP) message passing algorithm and convex optimization inspired researchers to apply graphical models concepts to CS theory to find faster solvers.

In order to connect CS theory with graphical models, first we model CS problem as a probabilistic inference problem. Figure 2.1 [subplot (a)] provides a block diagram representation for (1.3). It is assumed that the sparse source is resulted

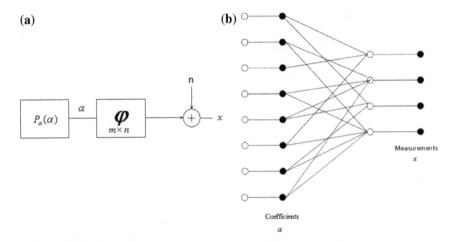

Fig. 2.1 (a) Probabilistic block diagram for CS and (b) corresponding factor graph

from sampling of a probability distribution $P_s(s)$. Sparse sources have been modeled with heavy-tailed distributions including Laplacian, Gaussian mixtures, generalized Gaussian, and Bernoulli Gaussian distributions in the literature [18]. The observation is resulted from the source via linear transformation, $A = \Phi\Psi$, followed by noise contamination. The goal is to estimate the source signal, either MAP or MMSE estimations, using the observed measurement, y. In this framework the original CS problem can be considered as a probabilistic inference problem. Exact MAP estimate can be computed for the problem [18] but unfortunately the solution involves heavier computational load compared to l_1-norm minimization methods. One idea is to use approximate inference algorithms such as BP to lessen the computational load. To do this end a graphical model must be assigned to the problem. The main idea for this purpose comes form error control coding area, where it is common to represent a parity check matrix by a biparitite graph. Analogously the block diagram in Fig. 2.1 [subplot (a)] can be represented by a biparitite factor graph as shown in Fig. 2.1 [subplot (b)]. There are two class of nodes in the factor graph: variable nodes (black) and constraint nodes (white). The edges connect variable nodes to constraint nodes. A constraint node models the dependencies that its neighboring variable nodes are subjected to. We have two types of constraint nodes; the first type imposes the probability distribution on source coefficients while the second type connects each coefficient node to a set of measurement variables that are used in computing that measurement. Having this factor graph, belief propagation can be employed to infer the probability distribution of the coefficients and consequently the MAP estimation for source signal.

In [18], the authors used belief propagation to infer the source signal. While their approach is interesting and the algorithm is much faster compared to general CS reconstruction algorithms, it poses a main limitation: to run BP, the authors assumed the sensing matrix to be sparse, which is not a realistic assumption in most CS applications. The reason for this assumption is that the implementation of BP in the general case is computationally intractable for dense graphs. Fortunately BP often admits acceptable solution for large, dense matrix when Gaussian approximation is used [19]. This property has led to generalization of approximate message passing algorithms for dense graphs. The key idea of generalized message passing algorithm (GMA) is to decompose the vector valued estimation problem into a sequence of scaler problems. This idea combined with the idea given in [18], has been used to generalize the compressive sensing algorithm via belief propagation for CS problems with dense sensing matrices. This class of algorithms are new compared to other classes and research is still going on to improve and generalize these algorithms to non-parametric cases, where we do not have prior information about the source distribution.

In this section a brief review on CS reconstruction algorithm was given. As stated in Chap. 1, one of the main limitations of applying CS to applications is at its reconstruction side. After about a decade of extensive research in this area, nowadays CS is well established and matured in terms of theory and analysis, but research is still going on to improve the current reconstruction algorithm in terms of computational and implementational complexity. Simple algorithms which can be implemented cheaply via electronic devices is crucial for this research area.

References

1. E.J. Candés, J. Romberg, T. Tao, Robust uncertainty principles: exact signal reconstruction from highly incomplete frequency information. IEEE Trans. Inf. Theor. **52**(2), 489–509 (2006)
2. Y. Tsaig, D.L. Donoho, Compressed sensing. IEEE Trans. Inf. Theor. **52**, 1289–1306 (2006)
3. E.J. Candès, The restricted isometry property and its implications for compressed sensing. C. R. Math. **346**(9–10), 589–592 (2008)
4. T. Strohmer, R.W.J. Heath, Grassmannian frames with applications to coding and communication. Appl. Comput. Harmonic Anal. **14**(3), 257–275 (2003)
5. S.A. Vavasis, *Elementary proof of the spherical section property for random matrices* (University of Waterloo, Waterloo,Technical report, 2009)
6. Y. Zhang.On theory of compressive sensing via l_1-minimization: Simple derivations and extensions. Technical Report TR08-11, Rice University, 2008
7. B. Kashin, V. Temlyakov, A remark on compressed sensing. Math. Notes **82**, 748–755 (2007)
8. A.W. Tucker, in *Proceedings of the second berkeley symposium on mathematical statistics and probability*, ed. by J. Neyman (University of California Press, Berkeley, 1951), pp. 481–492
9. S.S. Chen, D.L. Donoho, M.A. Saunders, Atomic decomposition by basis pursuit. SIAM Rev. **43**(1), 129–159 (2001)
10. E. van den Berg, M.P. Friedlander, Probing the pareto frontier for basis pursuit solutions. SIAM J. Sci. Comput. **31**(2), 890–912 (2008)
11. M.A.T. Figueiredo, R.D. Nowak, S.J. Wright, Gradient projection for sparse reconstruction: Application to compressed sensing and other inverse problems. IEEE J. Sel. Top. Sign. Proces. **1**(4), 586–597 (2007)
12. A. Beck, M. Teboulle, A fast iterative shrinkage-thresholding algorithm for linear inverse problems. SIAM J. Imaging Sci. **2**, 183–202 (2009)
13. C. Vonesch, M. Unser, A fast iterative thresholding algorithm for wavelet-regularized deconvolution, in *Proceedings of the SPIE Conference on Mathematical Imaging: Wavelet XII*, vol 6701, (2007) pp 67010D–1-67010D-5
14. Y. C. Pati, R. Rezaiifar, P. S. Krishnaprasad, Orthogonal matching pursuit: recursive function approximation with applications to wavelet decomposition, in *Proceedings of 27th Asilomar Conference on Signals Systems and Computers*, (1993) pp. 40–44
15. D. Needell, CoSaMP: Iterative signal recovery from incomplete and inaccurate samples. Commun. ACM **53**(12), 93–100 (2010)
16. G.H. Mohimani, M. Babaie-Zadeh, C. Jutten, A fast approach for overcomplete sparse decomposition based on smoothed l_0 norm. IEEE Trans. Sig. Proces. **57**(1), 289–301 (2009)
17. D. Needal, J. Tropp, Iterative signal recovery from incomplete and inaccurate samples. Appl. Comp. Harm. Anal. **26**, 301–321 (2008)
18. D. Baron, S. Sarvotham, R.G. Baraniuk, Bayesian compressive sensing via belief propagation. IEEE Trans. Sig. Proces. **58**(1), 269–280 (2010)
19. M. Bayati, A. Montanari, The dynamics of message passing on dense graphs, with applications to compressed sensing. IEEE Trans. Inf. Theor. **57**(2), 764–785 (2011)

Chapter 3
Compressed Sensing with Side Information on Feasible Region

In the literature the problem of compressed sensing in the presence of side information is studied. But, in most cases the side information are about the source itself, i.e. structure, probability distribution, etc. In this chapter, the problem of compressed sensing in the presence of side information about the feasible region is reviewed. We follow an approach similar to [1] to formulate the problem mathematically for a wider class. Next it is shown that uniqueness and stability results of CS still holds in this formulation. Finally, an efficient recovery algorithm is derived which incorporates the side information.

3.1 Formulation

Consider a general compressed sensing problem (1.3). Assume null(A) satisfies spherical section property with parameter Δ, consequently Theorems 2.3.1 and 2.3.2 hold for this problem. From linear algebra if $\mathbf{s}_1$ is a special solution to the system $A\mathbf{s} = \mathbf{y}$, then the feasible region for the optimization problem (1.3) would be:

$$F = \{\mathbf{s}_1 + \mathbf{s}_2 | A\mathbf{s}_2 = 0\} = \{\mathbf{s}_1\} + \text{null}(A). \tag{3.1}$$

In the current work we adopt FISTA as the reconstruction algorithm. To solve for the unique solution we start from an initial point and then search in the feasible region in (2.34). The size of this region depends on null(A) $\equiv \mathbb{R}^m$ $(rank(A))$ and intuitively we expect the bigger this space is, the harder is to solve the optimization problem. In other words when the feasible region is small then (2.34) converges faster to the solution of (1.4). Thus any side information about the feasible region is helpful.

Now consider cases that we have side information about the feasible region. For instance in the case of derivative compressed sensing (DCS) [1], where the source signal is a gradient field, the side information will result in $B\mathbf{s} = 0$ condition on the source signal ($B \in \mathbb{R}^{\frac{n}{2} \times n}$ and is resulted from inherent property of a gradient field.

M. Rostami, *Compressed Sensing with Side Information on the Feasible Region*,
SpringerBriefs in Electrical and Computer Engineering,
DOI: 10.1007/978-3-319-00366-5_3, © The Author(s) 2013

We will discuss this special case in more details in Chap. 4). A more general case may happen when we have side information as:

$$Bs = b, \tag{3.2}$$

where $B \in \mathbb{R}^{m' \times n}$ is a full rank matrix. Many constraints on a source can be formulated as (3.2). In such cases we have two types of information about the source. We call the first type, primary information, which is resulted through measurements $(As = y)$. The secondary information comes in hand through an inherent property of the source. Broadly speaking, we can assume we have a general inverse problem $(Bs = b)$ and we also have sparsity prior on the source, then we apply CS as a regularization method on this problem. Some problems in image/signal processing area such as image super-resolution, image impainting, and medical imaging can be modeled in this framework. We expect that if we incorporate this side information it somehow improves CS reconstruction. For instance we may be able to decrease the number of measurements for recovering the source with similar accuracy or some kind of robustness towards noise when the measurements are noisy.

Let $A' = \begin{bmatrix} A \\ B \end{bmatrix}$ ($A' \in \mathbb{R}^{(m+m') \times n}$ and is full rank matrix) and $y' = \begin{bmatrix} y \\ b \end{bmatrix}$ ($y' \in \mathbb{R}^{(m+m')}$). We have the following equivalent problem:

$$\hat{s} = \arg\min_{s} \|s\|_0 \quad \text{s.t.} \quad y' = A's. \tag{3.3}$$

Assume $m + m' \leq n$, in such cases the new problem is exactly in theform of a CS problem. We assume $m + m' \leq n$ so as to ensure an underdetermined system to deal with problem in CS framework. Now the question is: Does this problem have a unique solution? Can we still replace the l_0-norm with l_1-norm? To answer these questions, the new sensing matrix A' must be studied. In the next section we will show the answers to both questions are positive.

3.2 Uniqueness and Stability

Consider the optimization problem (3.3). Our assumption is that A has Δ-spherical section property, so (1.3) has a unique solution $\hat{s}$ and also we have l_1-l_0 equivalence. We show that adding the secondary condition will not violate the uniqueness, and furthermore solutions of (1.3) and (3.3) are equal.

Lemma 1 *The problem (3.3) has a unique solution $\hat{s}$ equivalent to the solution of (1.3). Furthermore l_1-l_0 equivalence holds for this problem, i.e.:*

$$\hat{s} = \arg\min_{s} \|s\|_1 \quad s.t. \quad y' = A's. \tag{3.4}$$

Proof 3 *The proof is simple. First we show A' has spherical section property with $\Delta' = (1 + \frac{m'}{m})\Delta$. Note $dim(null(A')) = n - (m + m')$ and:*

$$A's = 0 \rightarrow \begin{cases} As = 0 \\ Bs = 0 \end{cases} \rightarrow null(A') = null(A) \cap null(B), \quad (3.5)$$

thus $null(A') \subset null(A)$. Consequently $\forall s \in null(A') \subset null(A)$:

$$\frac{\|s\|_1}{\|s\|_2} \geq \sqrt{\frac{m}{\Delta}} = \sqrt{\frac{m + m'}{\Delta(1 + \frac{m'}{m})}}, \quad (3.6)$$

according to the definition of SSP, $null(A')$ has spherical section property with $\Delta = (1 + \frac{m'}{m})\Delta$.

  According to the assumption $\hat{s}$ is the solution of (1.3) and also satisfies $Bs = b$, $\hat{s}$ lays in the feasible region of (3.3). According to Theorem 2.3.1, $\hat{s}$ is a unique solution of (3.3), if $\|\hat{s}\|_0 \leq \frac{m+m'}{2\Delta'}$. Note $\|\hat{s}\|_0 \leq \frac{m}{2\Delta}$ since it is the unique solution of (1.3), then from Theorem 2.3.1:

$$\|\hat{s}\|_0 \leq \frac{m}{2\Delta} = \frac{m(m + m')}{2\Delta(m + m')} = \frac{m + m'}{2\Delta(1 + \frac{m'}{m})} = \frac{m + m'}{2\Delta'}, \quad (3.7)$$

which concludes the proof. Similarly we can conclude if the original primary CS problem has l_1-l_0 equivalence in Theorem 2.3.2, then (3.3) inherits this property. Also note that this unique solution satisfies $As = y$ and thus is equal to solution of (1.3).

This lemma states that we can add any side information in the form of (3.2) to our problem and this will not make the situation worse. This result is very intuitive and is expected but the Lemma also gives a mathematical justification. For source reconstruction we can use a general proposed CS reconstruction algorithm and find the unique solution of (3.3), however this may not be efficient enough. In the next section, a more efficient algorithm is proposed to solve (3.3).

3.3 Numerical Solution Algorithm

As explained, the problem that we formulated in Sect. 3.1 can be formulated by (3.3). We also expect some improvement if we use side information. In this section an efficient algorithm is derived for solving this problem. When Lemma 1 holds, (3.3) can be equivalently formulated as follows:

$$\hat{s} = \arg\min_{s} \mu \|s\|_1 + \|As - y\|_2^2 \quad s.t. \quad Bs = b, \quad (3.8)$$

now we have our original CS reconstruction problem constrained to the side information. To solve optimization problems in this form one can use operator splitting [2–4]. We will have a quick review on this method and then use it to solve (3.8).

3.3.1 Bregman Iteration and Operator Splitting

Consider the following optimization problem:

$$\min_{\mathbf{s}} J(\mathbf{s}) \quad \text{s.t.} \quad H(\mathbf{s}) = 0, \tag{3.9}$$

where H is a convex differentiable furcation while J is also convex but possibly non-differentiable functions. An efficient method to solve this type of problems is to use the Bregman iterations [2].

To proceed we need the definition of sub-gradient and Bregman distance.

Definition 6 *Let $J(\cdot) : \mathbb{R}^n \rightarrow \mathbb{R}^+$ be a convex and possibly non-differentiable function. The vector $\mathbf{p} \in \mathbb{R}^n$ is called a sub-gradient of J at point $\mathbf{w}_0$:*

$$\forall \mathbf{w} \in \mathbb{R}^n : J(\mathbf{w}) - J(\mathbf{w}_0) \leq \langle \mathbf{p}, \mathbf{w} - \mathbf{w}_0 \rangle. \tag{3.10}$$

Also, the set of all $\mathbf{p}$'s is called sub-differentiable of J at point $\mathbf{w}$ and is denoted by $\partial J(\mathbf{w}_0)$.

For a differentiable function, $\partial J(\mathbf{w}_0)$ reduces to a singleton which only contains the gradient vector, $\nabla J(\mathbf{w}_0)$. This concept extends the definition of gradient to convex but possibly non-differentiable functions. For instance sub-differentiable of $J(w) = |w|$ at the point $w = 0$ is the set $[-1, 1]$. Next we require definition of Bregman distance

Definition 7 *The Bregman distance of a convex function $J(\cdot) : \mathbb{R}^n \rightarrow \mathbb{R}^+$ between two points $\mathbf{s}$ and $\mathbf{w}$ is defined as:*

$$D_J^p(\mathbf{s}, \mathbf{w}) = J(\mathbf{s}) - J(\mathbf{w}) - \langle \mathbf{p}, \mathbf{s} - \mathbf{w} \rangle, \tag{3.11}$$

where $\mathbf{p}$ is a sub-gradient of J at $\mathbf{w}$.

Note (3.11) is not symmetric, thus Bregman distance is not a metric but somehow measures closeness of the two points.

Now back in our problem (3.9), this problem can be solved iteratively as follows:

$$\begin{cases} \mathbf{s}^{i+1} = \arg\min_{\mathbf{s}} D_J^{\mathbf{p}^i}(\mathbf{s}, \mathbf{s}^i) + \delta H(\mathbf{s}) \\ \mathbf{p}^{i+1} = \mathbf{p}^i - \delta \partial H(\mathbf{s}^{i+1}), \end{cases} \tag{3.12}$$

where $\delta \geq 0$ is a constant. It is shown in [4], that if the original problem (3.9) has a solution $\hat{\mathbf{s}}$, then through the iterations in (3.12), as $i \to \infty$ then $\mathbf{s}^i \to \hat{\mathbf{s}}$.

Now we apply this algorithm on (3.8), for which we can assume, $H(\mathbf{s}) = \frac{1}{2}\| B\mathbf{s} - \mathbf{b}\|_2^2$ and $J(s) = \mu\|\mathbf{s}\|_1 + \|A\mathbf{s} - \mathbf{y}\|_2^2$. This will reduce (3.12):

$$\begin{cases} (\mathbf{s}^{i+1}, \mathbf{b}^{i+1}) = \arg\min_{\mathbf{s},\mathbf{b}} \mu\|\mathbf{s}\|_1 + \frac{1}{2}\|A\mathbf{s} - \mathbf{y}\|_2^2 + \frac{\delta}{2}\| B\mathbf{s} - \mathbf{b} + \mathbf{p}^i \|_2^2 \\ \mathbf{p}^{i+1} = \mathbf{p}^i + \delta B\mathbf{s}^{i+1} - \mathbf{b}^{i+1}. \end{cases} \quad (3.13)$$

Note that the update step of the first equation in (3.13) has the format of a standard basis pursuit de-noising (BPDN) problem [5], which can be solved by a variety of optimization methods [6]. In the present paper, we used the FISTA algorithm of [7] due to the simplicity of its implementation as well as for its remarkable convergence properties. It should be noted that the algorithm does not require explicitly defining the matrices A and B. Only the *operations* of multiplication by these matrices and their transposes need to be known, which can be implemented in an implicit and computationally efficient manner. The main advantage of solving the problem using operator splitting is the much faster convergence of the thresholding algorithm.

Now equipped with some theoretical evidence and an efficient reconstruction algorithm we continue with some experimental study on advantageous prospect of our approach.

3.4 Experimental Study

To verify our analysis and algorithm, this section is devoted to experimental study on synthetics data, where as in the next Chapters we will focus on the practical applications of the developed method.

3.4.1 Source Model

For source simulation we used mixture of Gaussian model as the sparse source model:

$$\mathbf{s} \sim pN(0, \sigma_1) + (1 - p)N(0, \sigma_2), \quad (3.14)$$

where $N(0, \sigma_i)$ denotes a Gaussian distribution with zero mean and variance $= \sigma_i^2$, $\sigma_1 \ll \sigma_2$, and p is the parameter for a Bernoulli distribution. This model has been used to represent sparse signals in the literature [8, 9]. Although it is not a proper model for some applications, it is useful for our experimental study. Here, it is assumed that the source signal has two states. The first state corresponds to source elements with large values (non-zero elements) and the second state corresponds to elements with negligible value (approximately zero elements). The Bernoulli distribution parameter

p decides for each element, what state is active and controls the level of sparsity, and then each state is modeled via a Gaussian distribution. It must be taken into account that we only use this procedure for producing the source and assume the user does not have any information about the source probability distribution.

We also need a type of side information about the source that we can model in the form of (3.2). For this purpose we assumed that we have a prior information about the positions and values of some of the large value elements. This assumption is a good embed for testing the proposed method. We transform this information to the form of (3.2). An example makes the procedure clear. Assume we have a sparse source $s \in \mathbb{R}^{10}$. Assume we know that the second and the fourth elements are non-zero and both equal to 2, thus one concludes:

$$B = \begin{bmatrix} 0 & 1 & 0 & 0 & 0 & 0 & 0 & 0 & 0 & 0 \\ 0 & 0 & 0 & 1 & 0 & 0 & 0 & 0 & 0 & 0 \end{bmatrix} \tag{3.15}$$

and:

$$\mathbf{b} = \begin{bmatrix} 2 \\ 2 \end{bmatrix}. \tag{3.16}$$

In the general case for $B \in \mathbb{R}^{m' \times n}$, where m' is the number of the known non-zero elements, in each row we set B_{ij} related to the jth known non-zero element equal to one and the rest of the matrix entries equal to zero. Trivially $\mathbf{b}_j$ is equal to the jth known non-zero element. We continue with experiments in this framework.

3.4.2 Experiments

We set $n = 1,024$, $\sigma_1 = 0.1$, and $\sigma_2 = 10$ with sparsity level of $p = 0.1$ to generate the source signal through this subsection. This means that we have an approximately sparse source with about 100 large value elements. The sensing matrix was chosen as a random matrix with i.i.d Gaussian entries and applied to the source to produce the measurement vector $\mathbf{y}$. This selection is standard in the CS literature because these matrices satisfy both RIP and SSP. Based on our assumption, the positions and the values of a fraction of large value elements are known and one can form (3.2). Through the experiments, we assumed one fourth of the large value elements of the source are known ($m' \approx 25$) and $m = 300$, unless stated.

Figure 3.1a depicts an instance of the generated source in which the signal is presented versus time (index). Figure 3.1b depicts the reconstruction results for the classical CS. Visually, it can be seen some approximately zero elements are estimated larger than the real values and the quality of the reconstruction is poor. Figure 3.1c depicts the reconstruction results for the proposed method. As it can be detected visually our method outperforms the classical method, which is also confirmed numerically by calculating the signal-to-noise ration (SNR). The result confirms that the proposed algorithm works properly and is able to incorporate the side information.

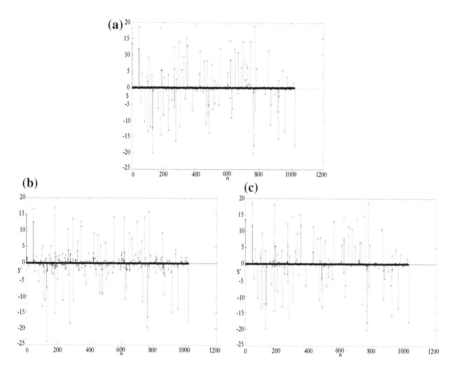

Fig. 3.1 **a** The source signal **b** reconstructed using the classical CS, SNR $= 14.4$ **c** reconstructed using the proposed method, SNR $= 26.0$

To analyze the algorithm two sets of experiments are done. First, we study the effect of the number of the known elements on the performance of the algorithm. Assume $0 < r \leq 1$ indicates the fraction of the known elements. Figure 3.2 depicts the proposed algorithm reconstruction quality (measured via SNR) versus r for $r \in (0.1, 1)$. As expected as we increase the side information, the quality of the reconstruction also improves such that for $r = 1$ we have near complete recovery.

In the second experiments we consider the effect of number of the measurements, m, on reconstruction quality. Figure 3.3 depicts output reconstruction SNR versus m for the classical CS and the proposed method. As expected, the reconstruction quality improves as the number of measurement increases for both methods. As it can be detected for large number of the measurements both methods are saturated and we have high SNR values. This is not surprising since when the number of the measurements are large enough we can reconstruct the source perfectly and the side information has negligible effect on the quality of the measurements. But for insufficient measurements, the side information becomes important and improves the quality of the reconstruction. This implies that the proposed method can be used to either improve the quality of the reconstruction or decrease the number of the required measurements without deteriorating the quality of the reconstruction. Overall, these experiments confirm the effectiveness of the proposed method. In the next chapters,

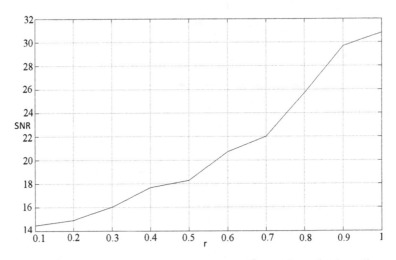

Fig. 3.2 Output SNR versus the fraction of known large value elements (r)

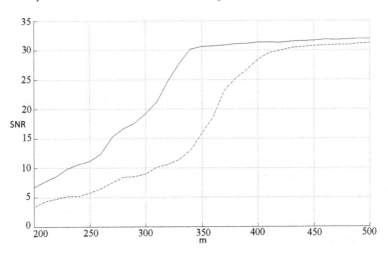

Fig. 3.3 SNR of the source reconstruction obtained with different methods as a function of m. Here, the dashed and solid lines correspond to the classical CS and the proposed CS method, respectively, and $r = 0.25$

this method has been applied to two practical examples: image deblurring in optical imaging [10] and surface reconstruction in the gradient field [11]. In both applications the source signals are gradient fields and the side information can be formulated as (3.2) as in [1]. Also, further analysis has been done through these applications.

References

1. M. Hosseini, O. Michailovich, Derivative compressive sampling with application to phase unwrapping. In *Proceedings of EUSIPCO*, Glasgow, UK, August, 2009
2. S. Osher, M. Burger, D. Goldfarb, J. Xu, W. Yin, An iterative regularization method for total variation-based image restoration. Simul **4**, 460–489 (2005)
3. W. Yin, S. Osher, D. Goldfarb, J. Darbon, Bregman iterative algorithms for ℓ_1-minimization with applications to compressed sensing. SIAM J. Imaging Sci. **1**(1), 143–168 (2008)
4. J. Cai, S. Osher, Z. Shen, Split Bregman methods and frame based image restoration. Multiscale Model. Simul. **8**(2), 337–369 (2009)
5. S.S. Chen, D.L. Donoho, Atomic decomposition by basis pursuit. SIAM J. Sci. Comput. **20**(1), 33–61 (1998)
6. I. Daubchies, M. Defrise, C.D. Mol, An iterative thresholding algorithm for linear inverse problems with sparsity constraint. Comm. Pure Appl. Math. **75**, 1412–1457 (2009)
7. A. Beck, M. Teboulle, A fast iterative shrinkage-thresholding algorithm for linear inverse problems. SIAM J. Imag. Sci. **2**, 183–202 (2009)
8. Z. Shi, H. Tang, Y. Tang, Blind source separation of more sources than mixtures using sparse mixture models. Pattern Recogn. Lett. **26**(16), 2491–2499 (2005)
9. S.A. Solla, T.K. Leen, K. Müller (eds.), *Advances in Neural Information Processing Systems*, vol. 12 (MIT Press, USA, 2000). [NIPS Conference, Denver, Colorado, USA, November 29–December 4, 1999]
10. M. Rostami, O.V. Michailovich, Z. Wang, Image deblurring using derivative compressed sensing for optical imaging application. IEEE Trans. Image Process. **21**(7), 3139–3149 (2012)
11. M. Rostami, O.V. Michailovich, Z. Wang, Gradient-based surface reconstruction using compressed sensing. In *19th IEEE International Conference on Image Processing*, Oralndo, U.S., 2012

Chapter 4
Application: Image Deblurring for Optical Imaging

The problem of reconstruction of digital images from their blurred and noisy measurements is unarguably one of the central problems in imaging sciences. Despite its ill-posed nature, this problem can often be solved in a unique and stable manner, provided appropriate assumptions on the nature of the images to be discovered. In this section, however, a more challenging setting is considered, in which accurate knowledge of the blurring operator is lacking, thereby transforming the reconstruction problem at hand into a problem of blind deconvolution [1, 2]. As a specific application, the current presentation focuses on reconstruction of short-exposure optical images measured through atmospheric turbulence. The latter is known to give rise to random aberrations in the optical wavefront, which are in turn translated into random variations of the point spread function (PSF) of the optical system in use. A standard way to track such variations involves using adaptive optics. For example, the Shack-Hartmann interferometer provides measurements of the optical wavefront through sensing its partial derivatives. In such a case, the accuracy of wavefront reconstruction is proportional to the number of lenslets used by the interferometer, and hence to its complexity. Accordingly, in this chapter, we show how to minimize the above complexity through reducing the number of the lenslets, while compensating for undersampling artifacts by means of derivative compressed sensing. Additionally, we provide empirical proof that the above simplification and its associated solution scheme result in image reconstructions, whose quality is comparable to the reconstructions obtained using conventional (dense) measurements of the optical wavefront.

4.1 Background

The necessity to recover digital images from their distorted and noisy observations is common for a variety of practical applications, with some specific examples including image denoising, super-resolution, image restoration, and watermarking, just to name

M. Rostami, *Compressed Sensing with Side Information on the Feasible Region*,
SpringerBriefs in Electrical and Computer Engineering,
DOI: 10.1007/978-3-319-00366-5_4, © The Author(s) 2013

a few [3–6]. In such cases, it is conventional to assume that the observed image v is obtained as a result of convolution of its original counterpart u with a point spread function[1] (PSF) i. To account for measurement inaccuracies, it is also standard to contaminate the convolution output with an additive noise term ν, which is usually assumed to be white and Gaussian. Thus, formally,

$$v = i * u + \nu. \tag{4.1}$$

While u and v can be regarded as general members of the signal space $\mathbb{L}_2(\Omega)$ of real-valued functions on $\Omega \subseteq \mathbb{R}^2$, the PSF i is normally a much smoother function, with effectively band-limited spectrum. As a result, the convolution with i has a destructive effect on the informational content of u, in which case v typically has a substantially reduced set of features with respect to u. This makes the problem of reconstruction of u from v a problem of significant practical importance [8].

Reconstruction of the original image u from v can be carried out within the framework of image deconvolution, which is a specific instance of a more general class of inverse problems [9]. Most of such methods are Bayesian in nature, in which case the information lost in the process of convolution with i is recovered by requiring the optimal solution to reside within a predefined functional class [10–12]. Thus, for example, in the case when u is known to be an image of bounded variation, the above regularization leads to the famous Rudin-Osher-Fatemi reconstruction scheme, in which u is estimated as a solution to the following optimization problem [13, 14]

$$\hat{u} = \arg\min_u \left\{ \frac{1}{2} \|u * i - v\|_2^2 + \alpha \int |\nabla u|\, dxdy \right\}, \tag{4.2}$$

where $\alpha > 0$ is the regularization parameter. It should be noted that, if the PSF obeys $\int i\, dxdy \neq 0$, the problem (4.2) is strictly convex and therefore admits a unique minimizer, which can be computed using a spectrum of available algorithms [13–17].

In some applications, the knowledge of the PSF may be lacking, which results in the necessity to recover the original image from its blurred and noisy observations alone. Such a reconstruction problem is commonly referred to as the problem of blind deconvolution [9]. In the present study, however, we follow the philosophy of *hybrid deconvolution* [18], whose main idea is to leverage any *partial* information on the PSF to improve the accuracy of image restoration. In particular, in the algorithm described in this chapter, such partial information is derived from incomplete observations of the partial derivatives of the phase of the generalized pupil function (GPF) of the optical system in use, as detailed below.

Optical imaging is unarguably the field of applied sciences from which the notion of image deconvolution has originated [19–21]. In particular, in short-exposure turbulent imaging [2], acquired images are blurred with a PSF, which depends on a spatial distribution of the atmospheric refraction index along the optical path connecting

---

[1] Note that, in optical imaging, this function is also referred to as an impulse transfer function [7].

an object of interest and the observer. Due to the effect of turbulence, the above distribution is random and time-dependent, which implies that the PSF i cannot be known in advance.

A standard way to overcome the above limitation is through the use of adaptive optics (AO) [22]. As will be shown later, the PSF of a short-exposure optical system is determined by its corresponding *generalized pupil function* (GPF) P, which can be expressed in a polar form as $P = A\,e^{j\phi}$. While, in practice, the amplitude A can be either measured through calibration or computed as a function of the aperture geometry, the phase ϕ accounts for turbulence-induced aberrations of the optical wavefront, and hence is generally unknown at any given experimental time. Fortunately, the phase ϕ turns out to be a measurable quantity, and this is where the tools of AO come into play. One of such tools is the Shack-Hartmann interferometer (SHI) (aka Shack-Hartmann wavefront sensor) [23–25], which allows direct measurement of the gradient of ϕ over a predefined grid of spatial coordinates. Subsequently, these measurements are converted into a useful estimate of ϕ through numerically solving an associated Poisson equation.

Among some other factors, the accuracy of phase reconstruction by the SHI depends on the size of its sampling grid, which is in turn defined by the number of lenslets composing the wavefront sensor of the interferometer (see below). Unfortunately, the grid size and the complexity (and, hence, the cost) of the interferometer tend to increase *pro rata*, which creates an obvious practical limitation. Accordingly, to overcome this problem, we propose to modify the construction of the SHI through reducing the number of its lenslets. Although the advantages of such a simplification are immediate to see, its main shortcoming is obvious as well: the smaller the number of lenslets is, the stronger is the effect of undersampling and aliasing. These artifacts, however, can be compensated for by subjecting the output of the simplified SHI to the derivative compressed sensing (DCS) algorithm of [26], which is a special case of the problem, studied in Chap. 3. As will be shown below, DCS is particularly suitable for reconstruction of ϕ from incomplete measurements of its partial derivatives. The resulting estimates of ϕ can be subsequently combined with A to yield an estimate of the PSF i, which can in turn be used by a deconvolution algorithm. Thus, the proposed method for estimation of the PSF i and subsequent deconvolution of u can be regarded as a hybrid deconvolution technique, which comes to simplify the design and complexity of the SHI on one hand, and to make the process of reconstruction of optical images as automatic as possible, on the other hand.

4.2 Technical Preliminaries

In short exposure imaging, due to phase aberrations in the optical wavefront induced by atmospheric turbulence, the PSF of an imaging system in use is generally unknown [2]. To better understand the setup under consideration, we first note that, in optical imaging, the PSF i is obtained from an amplitude spread function (ASF) h as $i :=$ $|h|^2$. The ASF, in turn, is defined in terms of a generalized pupil function (GPF) $P(x, y)$ and is given by [27]

$$h(\xi, \eta) = \frac{1}{\lambda_w z_i} \int_{-\infty}^{\infty} \int_{-\infty}^{\infty} P(x, y) e^{-j \frac{2\pi}{\lambda z_i} (x \xi + y \eta)} \, dx \, dy, \tag{4.3}$$

where z_i is the focal distance and λ_w is the optical wavelength. Being a complex-valued quantity, $P(x, y)$ can be represented in terms of its amplitude $A(x, y)$ and phase $\phi(x, y)$ as

$$P(x, y) = A(x, y) \, e^{j \phi(x, y)}. \tag{4.4}$$

Here, the GPF amplitude $A(x, y)$ (which is sometimes simply referred to as the aperture function) is normally a function of the aperture geometry. Thus, for instance, in the case of a circular aperture, $A(x, y)$ can be defined as [28]

$$A(r) = \begin{cases} 1, & \text{if } r \leq \frac{D}{2} \\ 0, & \text{otherwise} \end{cases} \tag{4.5}$$

where D denotes the pupil diameter. Thus, given $\phi(x, y)$, one could determine h and therefore i. Unfortunately, the phase $\phi(x, y)$ is influenced by the random effect of atmospheric turbulence, and as a result cannot be known ahead of time.

A standard way to overcome the uncertainty in $\phi(x, y)$ is to measure it using the tools of shearing interferometry, a particular example of which is the SHI [23, 29]. The latter is capable of sensing the partial derivatives of $\phi(x, y)$ over a predefined grid of spatial locations. In this case, an accurate reconstruction of $\phi(x, y)$ entails taking a fairly large number of the samples of $\nabla \phi(x, y)$, which is essential for minimizing the effect of aliasing on the estimation result [30]. Thus, in some applications, the number of sampling points (as defined by the number of SHI lenslets) reaches as many as a few thousands. It goes without saying that reducing the number of lenslets would have a positive impact on the SHI in terms of its cost and approachability. Alas, such a reduction is impossible without undersampling, which is likely to have a formidable effect on the overall quality of phase estimation.

To minimize the effect of phase undersampling, we exploit the DCS algorithm of [31]. The latter can be viewed as an extension of the conventional compressed sensing (CCS) scheme, in which the standard sparsity constraints are supplemented by additional constraints related to some intrinsic properties of partial derivatives. This "side information"—which are called the cross-derivative constraints—allows substantially improving the quality of reconstruction of $\phi(x, y)$, as compared to the case of CCS-based estimation.

4.2.1 Shack-Hartmann Interferometer

As it was mentioned the SHI can be used to measure the gradient $\nabla \phi(x, y)$ of the GPF phase $\phi(x, y)$, from which its values can be subsequently inferred. A standard approach to this reconstruction problem is to assume the unknown phase $\phi(x, y)$ to be expandable in terms of some basis functions $\{Z_k\}_{k=0}^{\infty}$, viz. [24]

$$\phi(x, y) = \sum_{k=0}^{\infty} a_k Z_k(x, y), \tag{4.6}$$

where the representation coefficients $\{a_k\}_{k=0}^{\infty}$ are supposed to be unique and stably computable. Note that, in this case, the datum of $\{a_k\}_{k=0}^{\infty}$ uniquely identifies $\phi(x, y)$, while the coefficients $\{a_k\}_{k=0}^{\infty}$ can be estimated due to the linearity of (4.6) which suggests

$$\nabla\phi(x, y) = \sum_{k=0}^{\infty} a_k \nabla Z_k(x, y). \tag{4.7}$$

In AO, it is conventional to define $\{Z_k\}_{k=0}^{\infty}$ to be Zernike polynomials (aka Zernike functions) [27]. These polynomials constitute an orthonormal basis in the space of square-integrable functions defined over the unit disk in $\mathbb{R}^2$. Zernike polynomials can be subdivided in two subsets of the even Z_n^m and odd Z_n^{-m} Zernike polynomials, which possess closed-form analytical definitions as given by

$$Z_n^m(\rho, \varphi) = R_n^m(\rho) \cos(m\,\varphi) \tag{4.8}$$

$$Z_n^{-m}(\rho, \varphi) = R_n^m(\rho) \sin(m\,\varphi), \tag{4.9}$$

where m and n are nonnegative integers with $n \geq m$, $0 \leq \varphi < 2\pi$ is the azimuthal angle, and $0 \leq \rho \leq 1$ is the radial distance. The radial polynomials R_n^m in (4.8) and (4.9) are defined as

$$R_n^m(\rho) = \sum_{k=0}^{(n-m)/2} \frac{(-1)^k (n-k)!}{k! ((n+m)/2 - k)! ((n-m)/2 - k)!} \rho^{n-2k}. \tag{4.10}$$

Note that, since the Zernike polynomials are defined using polar coordinates, it makes sense to re-express the phase ϕ and its gradient in the polar coordinate system as well. (Technically, this would amount to replacing x and y in (4.6), (4.7) by ρ and φ, respectively.) Moreover, due to the property of the Zernike polynomials to be an orthonormal basis, the representation coefficients $\{a_k\}_{k=0}^{\infty}$ in (4.6), (4.7) can be computed by orthogonal projection, namely

$$a_k = \int_0^{2\pi} \int_0^1 \phi(\rho, \varphi) Z_k(\rho, \varphi) \rho \, d\rho \, d\varphi. \tag{4.11}$$

In practice, however, $\phi(\rho, \varphi)$ is unknown and therefore the coefficients $\{a_k\}_{k=0}^{\infty}$ need to be estimated by other means. Thus, in the case of the SHI, the coefficients can be estimated from a finite set of discrete measurements of $\nabla\phi(\rho, \varphi)$.

The main function of the SHI is to acquire discrete measurements of $\nabla\phi$ by means of linearization. The linearization takes advantage of subdividing a (circular) aperture into rectangular blocks with their sides formed by a uniform rectangular

Fig. 4.1 An example of
a 10 × 10 SHI array on a
circular aperture. The *shading*
indicates those blocks (i.e.,
lenslets) which are rendered
active

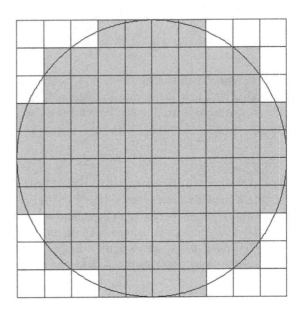

lattice. An example of such a subdivision is shown in Fig. 4.1 for the case of a 10×10 lattice grid. In general, the grid is assumed to be sufficiently fine to approximate ϕ by a linear function over the extent of a single block. This results in a piecewise linear approximation of ϕ, whose accuracy improves asymptotically when the lattice size goes to infinity. Formally, let $\Omega := \{(x, y) \in \mathbb{R}^2 \mid x^2 + y^2 \leq D^2\}$ be a circular aperture of radius D and $\mathcal{S} = \{(x, y) \in \mathbb{R}^2 \mid \max\{|x|, |y|\} \leq D\}$ be a square subset of $\mathbb{R}^2$ such that $\Omega \subset \mathcal{S}$. Then, for each polar coordinate $(\rho, \varphi) \in \Omega$ and an $N \times N$ grid of square blocks of size $2D/N \times 2D/N$, the phase ϕ can be expressed as

$$\phi(x, y) \approx ax + by + c, \tag{4.12}$$

for all (x, y) in a neighbourhood of $(\rho \cos \varphi, \rho \sin \varphi)$. The approximation in (4.12) suggests that

$$\nabla \phi(x, y) \approx (a, b)^T \tag{4.13}$$

where $(\cdot)^T$ denotes matrix transposition. While c in (4.12) can be derived from boundary conditions, coefficients a and b should be determined through direct measurements. To this end, the SHI is endowed with an array of small focusing lenses (i.e., lenslets), which are supported over each of the square blocks of the discrete grid, thereby forming a wavefront sensor. In the absence of phase aberrations, the focal points of the lenslets are spatially identified and registered using a high-resolution CCD detector, whose imaging plane is aligned with the focal plane of the sensor. Then, when the wavefront gets distorted by atmospheric turbulence, the focal points are dislocated towards new spatial positions, which can also be pinpointed by the same detector. The resulting displacements can be measured and subsequently related to the values of $\nabla \phi$ at corresponding points of the sampling grid.

To explain how the above procedure can be performed, additional notations are in order. Let Ω_d denote a finite set of spatial coordinates defined according to

$$
\begin{aligned}
\Omega_d := \Big\{ (x_d, y_d) \in \Omega \mid & \\
x_d = -D + \frac{2D}{N}\left(i + \frac{1}{2}\right), & \quad i = 0, 1, \dots, N - 1 \\
y_d = -D + \frac{2D}{N}\left(j + \frac{1}{2}\right), & \quad j = 0, 1, \dots, N - 1 \\
\text{and } x_d^2 + y_d^2 \leq D^2 \Big\}.
\end{aligned}
\tag{4.14}
$$

The set Ω_d can be thought of as a set of the spatial coordinates of the geometric centres of the SHI lenslets, restricted to the domain of its aperture Ω. Under the assumption of (4.12), one can then show [1] that the focal displacement $\Delta(x, y) = [\Delta_x(x, y), \Delta_y(x, y)]^T$ measured at some $(x, y) \in \Omega_d$ is related to the value of $\nabla\phi(x, y)$ according to

$$
\nabla\phi(x, y) \approx \frac{1}{F}\Delta\phi(x, y), \quad \forall (x, y) \in \Omega_d,
\tag{4.15}
$$

where F is the focal distance of the wavefront lenslets. An example of the above measurement setup is depicted in Fig. 4.2.

Now, provided a total of $M := \#\Omega_d$ ($\#\Omega_d$ denotes the cardinality of Ω_d) measurements of $\nabla\phi$ over Ω_d, one can approximate the coefficients $\{a_k\}_{k=1}^L$ of a truncated series expansion of ϕ as a solution to the least-square minimization problem given by

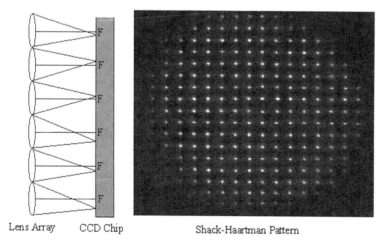

Lens Array   CCD Chip   Shack-Haartman Pattern

Fig. 4.2 Basic structure of the SHI and a resulting pattern of the focal points

$$\min_{\{a_k\}} \sum_{(x,y)\in\Omega_d} \| \sum_{k=0}^{L} a_k \nabla Z_k(x,y) - F^{-1}\Delta(x,y)\|_2^2, \qquad (4.16)$$

subject to appropriate boundary conditions. It is worthwhile noting that (4.16) can be rewritten in a vector-matrix form as

$$\min_{\mathbf{a}} \|\mathbf{Z}\mathbf{a} - \mathbf{d}\|_2^2, \quad \text{s.t. } \mathbf{a} \geq 0, \qquad (4.17)$$

where $\mathbf{Z}$ is a $2M \times L+1$ matrix of discrete values of the partial derivatives of the Zernike polynomials, $\mathbf{d}$ is a measurement (column) vector of length $2M$, and $\mathbf{a} = [a_0, a_1, \ldots, a_L]^T$ is a vector of the representation coefficients of ϕ. The constraint $\mathbf{a} \geq 0$ in (4.17) is optional and may be used to further regularize the solution by forcing $\mathbf{a}$ to belong to some convex set $\mathcal{K}_{\geq}$. Thus, for example, when the set coincides with the whole $\mathbb{R}^{L+1}$, the solution to (4.17) is given by

$$\mathbf{a} = \mathbf{Z}^{\#}\mathbf{d}, \qquad (4.18)$$

where $\mathbf{Z}^{\#}$ denotes the Moore-Penrose pseudo-inverse of $\mathbf{Z}$, whose definition is unique and stable as long as the row-rank of $\mathbf{Z}$ is greater or equal to $L+1$ (hence suggesting that $2M \geq L+1$). Having estimated $\mathbf{a}$, the phase ϕ can be approximated as

$$\phi(\rho, \varphi) \approx \sum_{k=0}^{L} a_k Z_k(\rho, \varphi). \qquad (4.19)$$

A higher accuracy of phase estimation requires using higher-order Zernike polynomials, which in turn necessitates a proportional increase in the number of wavefront lenses. Moreover, as required by the linearization procedure in the SHI, the lenses have to be of a relatively small sizes (sometimes, on the order of a few microns), which may lead to the use of a few thousand lenses per one interferometer. Accordingly, to simplify the construction and to reduce the cost of SHIs, we propose to reduce the number of wavefront lenslets, while compensating for the induced information loss through the use of DCS, which is detailed next.

4.3 Point Spread Function Estimation via Compressive Sampling

We now apply the proposed algorithm on this problem. First we show the side data, a source signal is a gradient field, can be transformed to (3.2) and then provide experiments that confirms that we can take advantage of the proposed scheme to improve the quality of image deblurring.

4.3.1 Derivative Compressed Sensing

Let the partial derivatives of ϕ evaluated at the points of set Ω_d be column-stacked into vectors $\mathbf{f}_x$ and $\mathbf{f}_y$ of length $M = \#\Omega_d$. In what follows, the partial derivatives $\mathbf{f}_x$ and $\mathbf{f}_y$ are assumed to be sparsely representable by an orthonormal basis in $\mathbb{R}^M$. Representing such a basis by an $M \times M$ unitary matrix W, the above assumption suggests the existence of two *sparse* vectors $\mathbf{c}_x$ and $\mathbf{c}_y$ such that $\mathbf{f}_x = W\mathbf{c}_x$ and $\mathbf{f}_y = W\mathbf{c}_y$. In the experimental studies of this section, the matrix W is constructed using the nearly symmetric orthogonal wavelets of I. Daubechies having five vanishing moments [32].

The proposed simplification of the SHI amounts to reducing the number of wave-front lenslets. Formally, such a reduction can be described by two $n \times M$ sub-sampling matrices Ψ_x and Ψ_y, where $n < M$. Specifically, let $\mathbf{b}_x := \Psi_x \mathbf{f}_x$ and $\mathbf{b}_y := \Psi_y \mathbf{f}_y$ be incomplete (partial) observations of $\mathbf{f}_x$ and $\mathbf{f}_y$, respectively. Then, based on the theoretical guarantees of classical CS, the vectors $\mathbf{f}_x$ and $\mathbf{f}_y$ of the partial derivatives of ϕ can be approximated by $W\mathbf{c}_x^*$ and $W\mathbf{c}_y^*$, respectively, where $\mathbf{c}_x^*$ and $\mathbf{c}_y^*$ are obtained as

$$\mathbf{c}_x^* = \arg\min_{\mathbf{c}_x'} \left\{ \frac{1}{2} \| \Psi_x W \mathbf{c}_x' - \mathbf{b}_x \|_2^2 + \lambda_x \|\mathbf{c}_x'\|_1 \right\} \qquad (4.20)$$

and

$$\mathbf{c}_y^* = \arg\min_{\mathbf{c}_y'} \left\{ \frac{1}{2} \| \Psi_y W \mathbf{c}_y' - \mathbf{b}_y \|_2^2 + \lambda_y \|\mathbf{c}_y'\|_1 \right\} \qquad (4.21)$$

for some $\lambda_x, \lambda_y > 0$. Moreover, in the case when $\lambda_x = \lambda_y := \lambda$, computing the above estimates can be combined into a single optimization problem. Specifically, let $\mathbf{c} = [\mathbf{c}_x, \mathbf{c}_y]^T$, $\mathbf{b} = [\mathbf{b}_x, \mathbf{b}_y]^T$, and $A = \text{diag}\{\Psi_x W, \Psi_y W\} \in \mathbb{R}^{2n \times 2M}$. Then,

$$\mathbf{c}^* = \arg\min_{\mathbf{c}'} \left\{ \frac{1}{2} \| A\mathbf{c}' - \mathbf{b} \|_2^2 + \lambda \|\mathbf{c}'\|_1 \right\}. \qquad (4.22)$$

In this form, the problem (4.22) is identical to (1.4), in which case it can be solved by a variety of available tools of convex optimization [33, 34].

The DCS algorithm augments classical CS by subjecting the minimization in (4.22) to an additional constraint which stems from the fact that [7]

$$\frac{\partial^2 \phi}{\partial x \, \partial y} = \frac{\partial^2 \phi}{\partial y \, \partial x}, \qquad (4.23)$$

which is valid for all twice continuously differentiable functions ϕ. Thus, in the discrete setting, the above condition can be expressed using two partial differences matrices D_x and D_y, in which case it reads

$$D_x \mathbf{f}_y = D_y \mathbf{f}_x. \qquad (4.24)$$

To further simplify the notations, let T_x and T_y be two coordinate-projection matrices, which map the composite vector $\mathbf{c}$ into $\mathbf{c}_x$ and $\mathbf{c}_y$ according to $T_x\mathbf{c} = \mathbf{c}_x$ and $T_y\mathbf{c} = \mathbf{c}_y$, respectively. Then (4.24) can be re-expressed in terms of $\mathbf{c}$ as

$$D_y W T_x \mathbf{c} = D_x W T_y \mathbf{c} \tag{4.25}$$

or, equivalently,

$$B\mathbf{c} = 0, \tag{4.26}$$

where $B := D_y W T_x - D_x W T_y$. Consequently, with the addition of the cross-derivative constraint (4.26), DCS solves the constrained minimization problem given by

$$\mathbf{c}^* = \arg\min_{\mathbf{c}'} \left\{ \frac{1}{2}\|A\mathbf{c}' - \mathbf{b}\|_2^2 + \lambda\|\mathbf{c}'\|_1 \right\}, \tag{4.27}$$

$$\text{s.t. } B\mathbf{c}' = 0.$$

The problem (4.27) is an instance of (3.8) and can be solved through the sequence of iterations produced by

$$\begin{cases} \mathbf{c}^{(t+1)} = \arg\min_{\mathbf{c}'} \left\{ \frac{1}{2}\|A\mathbf{c}' - \mathbf{b}\|_2^2 \right. \\ \qquad\qquad \left. +\lambda\|\mathbf{c}'\|_1 + \frac{\delta}{2}\|B\mathbf{c}' + p^{(t)}\|_2^2 \right\} \\ p^{(t+1)} = p^{(t)} + \delta B\mathbf{c}^{(t+1)}, \end{cases} \tag{4.28}$$

where $p^{(t)}$ is a vector of Bregman variables (or, equivalently, augmented Lagrange multipliers) and $\delta > 0$ is a user-defined parameter.[2]

Once an optimal $\mathbf{c}^*$ is recovered, it can be used to estimate the noise-free versions of $\mathbf{f}_x$ and $\mathbf{f}_y$ as $W T_x \mathbf{c}^*$ and $W T_y \mathbf{c}^*$, respectively. These estimates can be subsequently passed on to the fitting procedure to recover the values of ϕ, which, in combination with a known aperture function A, provide an estimate of the PSF i as an inverse discrete Fourier transform of the autocorrelation of $P = A e^{J\phi}$. Algorithm 1 below summarizes our method of estimation of the PSF.

The estimated PSF can be used to recover the original image u from v through the process of deconvolution as explained in the section that follows.

4.3.2 Deconvolution

The acquisition model 4.1 can be rewritten in an equivalent operator form as given by

---

[2] In this work, we use $\delta = 0.5$.

Algorithm 1: PSF estimation via DCS
___

1. *Data:* $\mathbf{b}_x$, $\mathbf{b}_y$, and $\lambda > 0$

2. *Initialization:* For a given transform matrix W and matrices/operators Ψ_x, Ψ_y, D_x, D_y, T_x and T_y, preset the procedures of multiplication by A, A^T, B and B^T.

3. *Phase recovery:* Starting with an arbitrary $\mathbf{c}^{(0)}$ and $p^{(0)} = 0$, iterate (4.28) until convergence to result in an optimal $\mathbf{c}^*$. Use the estimated (full) partial derivatives $W T_x \mathbf{c}^*$ and $W T_y \mathbf{c}^*$ to recover the values of ϕ over Ω.

4. *PSF estimation:* Using a known aperture function A, compute the inverse Fourier transform of $P = A e^{J\phi}$ to result in a corresponding ASF h. Estimate the PSF i as $i = |h|^2$.

___

$$v = \mathcal{H}\{u\} + \nu, \tag{4.29}$$

where $\mathcal{H}$ denote the operator of convolution with the estimated PSF i. Note that, in this case, the noise term ν accounts for both measurement noise as well as the inaccuracies related to estimation error in i.

The deconvolution problem of finding a useful approximation of u given its distorted measurement v can be addressed in many way, using a multitude of different techniques [35–39]. In this work, we use the ROF model and recover a regularized approximation of the original image u as

$$u^* = \arg\min_u \left\{ \frac{1}{2} \|\mathcal{H}\{u\} - v\|_2^2 + \gamma \|u\|_{TV} \right\}, \tag{4.30}$$

where $\|u\|_{TV} = \int \int |\nabla u| \, dx \, dy$ denotes the total variation (TV) semi-norm of u.

The minimization problem in (4.30) can be solved using a magnitude of possible approaches. One particularly efficient way to solve (4.30) is to substitute a direct minimization of the cost function in (4.30) by recursively minimizing a sequence of its local quadratic majorizers [38]. In this case, the optimal solution u^* can be obtained as the stationary point of a sequence of intermediate solutions produced by

$$\begin{cases} w^{(t)} = u^{(t)} + \mu \mathcal{H}^* \left\{ v - \mathcal{H}\{u^{(t)}\} \right\} \\ u^{(t+1)} = \arg\min_u \left\{ \frac{1}{2} \|u - w^{(t)}\|_2^2 + \gamma \|u\|_{TV} \right\}, \end{cases} \tag{4.31}$$

where $\mathcal{H}^*$ is the adjoint of $\mathcal{H}$ and μ is chosen to satisfy $\mu > \|\mathcal{H}^*\mathcal{H}\|$. In this work, the TV denoising at the second step of (4.31) has been performed using the fixed-point algorithm of Chambolle [14]. The convergence of (4.31) can be further improved by using the same FISTA algorithm of [38]. The resulting procedure is summarized below in Algorithm 2.

Algorithm 2: TV deconvolution using FISTA

---

1. *Initialize:* Select an initial value $u^{(0)}$; set $y^{(0)} = u^{(0)}$ and $\tau^{(0)} = 1$

2. *Repeat until convergence:*

- $w^{(t)} = y^{(t)} + \mu \mathcal{H}^* \left\{ v - \mathcal{H}\{y^{(t)}\} \right\}$
- $u^{(t+1)} = \arg\min_u \left\{ \frac{1}{2} \|u - w^{(t)}\|_2^2 + \gamma \|u\|_{TV} \right\}$
- $\tau^{(t+1)} = 0.5 \left(1 + \sqrt{1 + 4(\tau^{(t)})^2} \right)$
- $y^{(t+1)} = u^{(t+1)} + (\tau^{(t)}/\tau^{(t+1)})(u^{(t+1)} - u^{(t)})$

In summary, Algorithms 1 and 2 represent the essence of the proposed algorithm for hybrid deconvolution of short-exposure optical images. Next, experimental results are provided which further support the value and applicability of the proposed methodology.

4.4 Experiments

To demonstrate the viability of the proposed approach, its performance has been compared against two reference methods. The first reference method used a dense sampling of the phase (as it would have been the case with a conventional design of the SHI), thereby eliminating the need for a CS-based phase reconstruction. The resulting method is referred below to as the dense sampling (DS) approach. Second, to assess the importance of incorporation of the cross-derivative constraints, we have used both classical CS and DCS for phase recovery. In what follows, comparative results for phase estimation and subsequent deconvolution are provided for all the above methods.

4.4.1 Phase Recovery

To assess the performance of the proposed and reference methods under controllable conditions, simulation data was used. The random nature of atmospheric turbulence necessitates the use of statistical methods to model its effect on a wavefront propagation. Specifically, in this work, the effect of atmospheric turbulence was simulated based on the modified Von Karman model [40]. This model is derived based on Kolmogorov's theory of turbulence which models atmospheric turbulence using temperature fluctuations [41]. In particular, under some general assumptions on the velocity of turbulent medium and the distribution of its refraction index, the Von Karman power spectrum density is given by

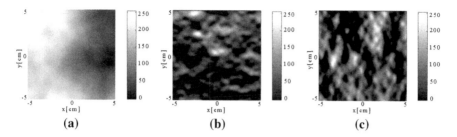

Fig. 4.3 An example of a simulated phase ϕ (a) along with its partial derivatives w.r.t. x (b) and y (c)

$$Q(t) = 0.033\, C_n^2\, \frac{e^{(-t^2/t_m^2)}}{(t^2 + t_0^2)^{11/6}}, \tag{4.32}$$

where C_n^2 is the refractive-index and t_m, t_0 are chosen to match the high frequency and low frequency behaviour of turbulence. The model of (4.32) can be used to generate random realizations of the GPF phase, as described, e.g., in [2].

A typical example of the GPF phase ϕ is shown in subplot (a) of Fig. 4.3. In this case, the size of the phase screen was set to be equal to 10×10 cm, while the sampling was performed over a 128×128 uniform grid (which would have corresponded to the use of 16384 lenslets of a SHI). The corresponding values of the (discretized) partial derivatives $\partial\phi/\partial x$ and $\partial\phi/\partial y$ are shown in subplots (b) and (c) of Fig. 4.3, respectively.

The subsampling matrices Ψ_x and Ψ_y were obtained from an identity matrix I through a random subsampling of its rows by a factor resulting in a required compression ratio r. To sparsely represent the partial derivatives of ϕ, W was defined to correspond to a four-level orthogonal wavelet transform using the nearly symmetric wavelets of I. Daubechies with five vanishing moments [32] and periodic boundary condition.

To demonstrate the value of using the cross derivative constraint for phase reconstruction, the classical CS and DCS algorithms have been compared in terms of the mean squared errors (MSE) of their corresponding phase estimates. The results of this comparison are summarized in Fig. 4.4 for different compression ratios (or, equivalently, (sub)sampling densities) and SNR = 40 dB.

As expected, one can see that DCS results in lower values of MSE as compared to classical CS, which implies a higher accuracy of phase reconstruction. Moreover, the difference in the performances of classical CS and DCS appears to be more significant for lower sampling rates, while both algorithms tend to perform similarly when the sampling density approaches the DS case. Specifically, for the sampling density of $r = 0.3$, DCS results in a ten times smaller value of MSE as compared to the case of classical CS, whereas both algorithms have comparable performance for $r = 0.83$. This result suggests that, at higher compression rates, DCS is likely to result in more accurate reconstructions of the GPF phase as compared to the case of classical CS.

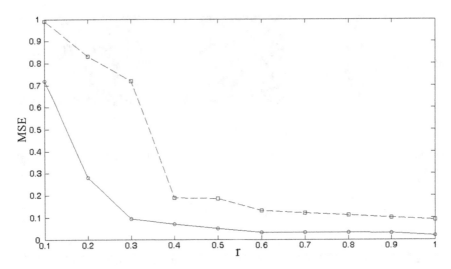

Fig. 4.4 MSE of phase reconstruction obtained with different methods as a function of r. Here, the *dashed* and *solid lines* correspond to classical CS and DCS, respectively, and SNR is equal to 40 dB

A number of typical reconstruction results are shown in Fig. 4.5, whose left and right subplots depict the phase estimates obtained using the classical CS and DCS algorithms, respectively, for the case of $r = 0.5$. The error maps of the two estimates are shown in subplot (c) and (d) of the same figure, which allows us to see the difference in the performance of these methods more clearly. Also, a close comparison with the original phase (as shown in subplot (a) of Fig. 4.3) reveals that DCS provides a more accurate recovery of the original ϕ, which further supports the value of using the cross-derivative constraints. In fact, exploiting these constraints effectively amounts to using additional "measurements", which are ignored in the case of classical CS.

As an additional comparison, Fig. 4.6 illustrates the convergence of the MSE as a function of the number of iterations, for both classical CS and DCS algorithms. One can see that DCS results in a substantially faster convergence as compared to classical CS. This behaviour could be explained by considering the cross-derivative constraints exploited by DCS to be effectively equivalent to noise-free measurements. To further investigate this argument, Fig. 4.7 compares the convergence of the cross-derivative fidelity term $\|D_y f_x - D_x f_y\|^2$ for both methods under comparison. One can see that, in the case of DCS, this term converges considerably faster than in the case of classical CS, which improves to the overall speed of convergence of DCS, making it superior to that of classical CS.

To investigate the robustness of the compared algorithms towards measurement noises, their performances have been compared for a range of SNR values. The results of this comparison are summarized in Fig. 4.8. Since the cross-derivative constraints exploited by DCS effectively restrict the feasibility region for an optimal solution, the algorithm exhibits an improved robustness to the effect of additive noise as compared

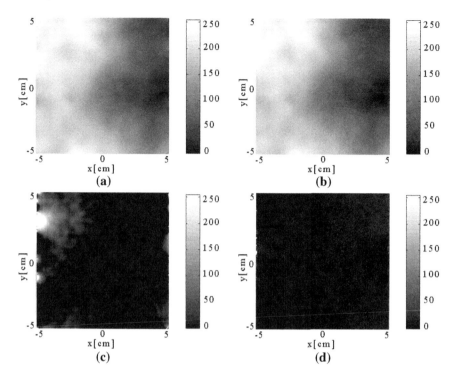

Fig. 4.5 **a** Phase reconstructed obtained by means of classical CS for SNR = 40 dB and $r = 0.5$; **b** Phase reconstructed obtained by means of DCS for the same values of SNR and r; **c** and **d** Corresponding error maps for classical CS and DCS

to the case of classical CS. This fact represents another advantage of incorporating the cross-derivative constraints in the process of phase recovery.

From the viewpoint of statistical estimation theory, the data fidelity terms in (4.27) suggests a Gaussian noise model, which may not be natural for all optical systems. In fact, this is the Poisson noise model, which is considered to be a more standard one in optical imagery. It turns out, however, that the use of the cross-derivative constraints by DCS makes it robust towards the inconsistency in noise modeling. This argument is supported by the results of Fig. 4.9, which summarizes the values of MSE obtained by classical CS and DCS reconstructions for different levels of Poisson noise. One can see that, in this case, the MSE values are comparable to the Gaussian case, while being substantially smaller in comparison to the CCS-based reconstruction.

It should be taken into account that, although the shape of ϕ does not change the energy of the PSF i, it plays a crucial role in determining its spatial behaviour. In the section that follows, it will be shown that even small inaccuracies in reconstruction of ϕ could be translated into dramatic difference in the quality of image deconvolution.

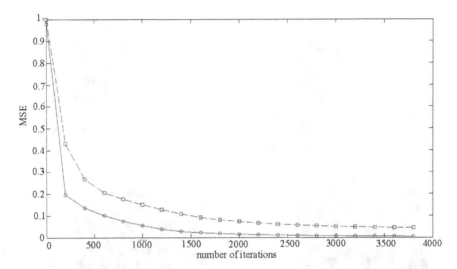

Fig. 4.6 Convergence analysis of phase reconstruction obtained with different methods as a function of iterations. Here, the *dashed* and *solid lines* correspond to classical CS and DCS, respectively, $SNR = 40$, and $r = 0.5$

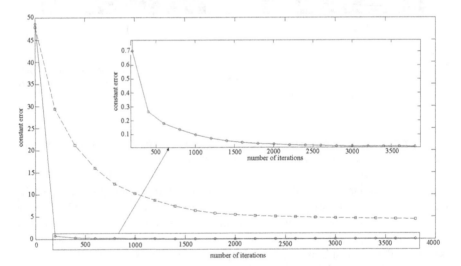

Fig. 4.7 Convergence analysis of derivative constraint obtained with different methods as a function of iterations. Here, the *dashed* and *solid lines* correspond to classical CS and DCS, respectively, $SNR = 40$, and $r = 0.5$

4.4.2 Deblurring

As a next step, the phase estimates obtained using the CCS- and DCS-based methods for $r = 0.5$ were combined with the aperture function A to result in their respective estimates of the PSF i. These estimates were subsequently used to deconvolve a

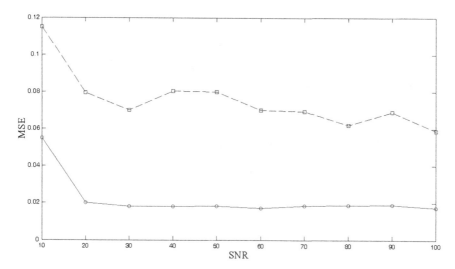

Fig. 4.8 MSE of phase reconstruction obtained with different methods as a function of SNR. Here, the *dashed* and *solid lines* correspond to classical CS and DCS, respectively, and $r = 0.5$

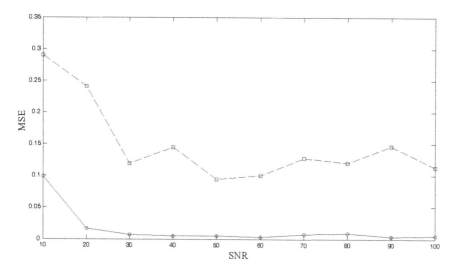

Fig. 4.9 MSE of phase reconstruction obtained with different methods as a function of SNR where the noise model is Poisson. Here, the *dashed* and *solid lines* correspond to classical CS and DCS, respectively, and $r = 0.5$

number of test images such as "Satellite", "Saturn", "Moon" and "Galaxy". All the test images were blurred with an original PSF, followed by their contamination with additive Gaussian noise of different levels which is controlled by the variance of noise distribution. As an example, Fig. 4.10 shows the "Satellite" image [subplot (a)] along with its blurred and noisy version [subplot (b)].

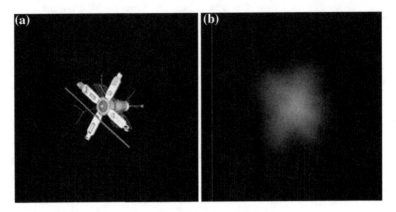

Fig. 4.10 Satellite image (**a**) and its blurred and noisy version (**b**)

Table 4.1 SSIM and PSNR comparisons of phase recovery results

Image	PSNR comparison (dB)					SSIM comparison			
	Noise std	Blurred	DS	CS	DCS	Blurred	DS	CS	DCS
	10^{-5}	14.06	27.97	17.06	27.42	0.200	0.730	0.349	0.674
Satellite	0.001	14.06	27.75	16.93	27.22	0.200	0.720	0.344	0.667
	0.003	14.06	25.97	16.54	25.56	0.199	0.554	0.306	0.519
	0.005	14.05	22.43	15.63	22.22	0.197	0.269	0.206	0.263
	10^{-5}	17.78	31.49	23.42	31.02	0.226	0.688	0.424	0.656
Saturn	0.001	17.78	31.08	23.38	30.65	0.226	0.66	0.416	0.641
	0.003	17.78	28.50	22.80	28.30	0.226	0.506	0.348	0.483
	0.005	17.78	23.89	20.55	23.72	0.175	0.228	0.212	0.223
	10^{-5}	19.98	25.06	22.36	25.00	0.512	0.645	0.539	0.643
Moon	0.001	19.97	25.04	22.38	24.99	0.512	0.642	0.538	0.64
	0.003	19.97	24.83	22.30	24.78	0.509	0.607	0.493	0.604
	0.005	19.97	21.76	19.73	21.73	0.504	0.552	0.488	0.549
	10^{-5}	18.79	23.58	21.16	23.52	0.257	0.493	0.348	0.490
Galaxy	0.001	18.79	23.60	21.12	23.54	0.257	0.495	0.347	0.491
	0.003	18.78	23.38	20.64	23.32	0.257	0.501	0.326	0.501
	0.005	18.78	20.93	18.46	20.86	0.254	0.397	0.224	0.393

Using the PSF estimates, the deconvolution was carried out using the method
detailed in [14]. For the sake of comparison, the deconvolution was also performed
using the PSF recovered from dense sampling (DS) of ϕ. Note that this reconstruction
is expected to have the best accuracy, since it neither involves undersampling nor
requires a CS-based phase estimation. All the deconvolved images have been compared with their original counterparts in terms of PSNR as well as of the structural
similarity index (SSIM) of [42], which is believed to be a better indicator of perceptual
image quality [43]. The resulting values of the comparison metrics are summarized
in Table 4.1, while Fig. 4.11 shows the deconvolution results produced by the CCS-
and DCS-based methods.

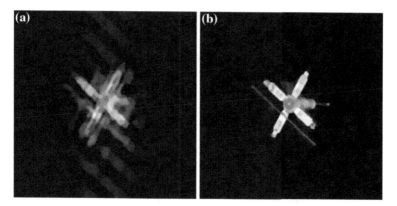

Fig. 4.11 a Image estimate obtained with the CCS-based method for phase recovery (SSIM = 0.781). **b** Image estimate obtained with the DCS-based method for phase recovery (SSIM = 0.917)

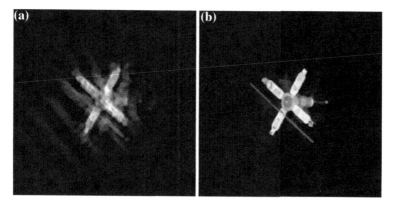

Fig. 4.12 a Image estimate obtained with the CCS-based method for phase recovery (SSIM = 0.732). **b** Image estimate obtained with the DCS-based method for phase recovery (SSIM = 0.888) where the noise model is assumed to be Poisson

The above results demonstrate the importance of accurate phase recovery, where even a relatively small phase error can have a dramatic effect on the quality of image deconvolution. Under such conditions, the proposed method produces image reconstructions of a superior quality as compared to the case of classical CS. Moreover, comparing the results of Table 4.1, one can see that DS only slightly outperforms DCS in terms of PSNR and SSIM, while in many practical cases, the difference between the performances of these methods are hard to detect visually.

Finally, Fig. 4.12 shows the results of CCS-based and DCS-based image reconstruction for the case of Poisson noise contamination. A close comparison of these results reveals a noticeable degradation in the performance of the CCS-based algorithm, while the DCS-based results are virtually indistinguishable from those obtained in the Gaussian case.

4.5 Summary

In this chapter, the applicability of the proposed scheme to the practical problem of image deblurring in optical imaging was studied. It was shown that, in the presence of atmospheric turbulence, the phase ϕ of the GPF $P = A\,e^{j\phi}$ is a random function, which needs to be measured using adaptive optics. To simplify the complexity of the latter, a CS-based approach was proposed. As opposed to classical CS, however, the proposed method performs phase reconstruction subject to an additional constraint, which stems from the property of $\nabla\phi$ to be a potential field. The DCS algorithm has been shown to yield phase estimates of substantially better quality as compared to the case of classical CS. our main focus has been on simplifying the structure of the SHI through reducing the number of its wavefront lenslets, while compensating for the effect of undersampling by means of DCS. The resulting phase estimates were used to recover their associated PSF, which was subsequently used for image deconvolution. It was shown that the DCS-based estimation of ϕ with $r = 0.3$ results in image reconstructions of the quality comparable to that of DS, while substantially outperforming the results obtained with classical CS.

References

1. J. Primot, G. Rousset, J.C. Fontanella, Deconvolution from wave-front sensing: A new technique for compensating turbulence-degraded images. J. Opt. Soc. Am. **7**, 1598–1608 (1990)
2. M.C. Roggemann, B.M. Welsh, *Imaging Through Turbulence* (CRC Press, Boca Raton, 1996)
3. J. Yang, J. Wright, T.S. Huang, Y. Ma, Image super-resolution via sparse representation. IEEE Trans. Image Processing **19**, 2861–2873 (2010)
4. M. Elad, M. Aharon, Image denoising via sparse and redundant representations over learned dictionaries. IEEE Trans. Image Processing **17**, 3736–3745 (2006)
5. I.J. Mairal, G. Sapiro, M. Elad, Learning multiscale sparse representations for image and video restoration. Multiscale Modeling and Simulation **7**, 214–241 (2008)
6. R.T. Paul, Review of robust video watermarking techniques. IJCA Special Issue on Computational Science **3**, 90–95 (2011)
7. G.D. Boreman, *Modulation Transfer Function in Optical and Electro-Optical Systems* (SPIE Optical Engineering Press, Bellingham, Washington, 2001)
8. R.T. Paul, Blind deconvolution via cumulant extrema. IEEE Signal Processing Magazine **3**, 24–42 (1996)
9. D. Kundur, D. Hatzinakos, Blind image deconvolution. IEEE Signal Processing Magazine **3**, 43–64 (1996)
10. J.K. Kauppinen, D.J. Moffatt, H.H. Mantsch, D.G. Cameron, Fourier self-deconvolution: A method for resolving intrinsically overlapped bands. Applied Spectroscopy **35**, 271–276 (1981)
11. S. Geman and D. Geman. Stochastic relaxation, Gibbs distribution and the Bayesian restoration of images. IEEE Trans. Pattern Analysis and Machine Intelligence, PAMI-6:721–741, 1984
12. T. Poggio, V. Torre, C. Koch, Computational vision and regularization theory. Nature **317**, 314–319 (1985)
13. L.I. Rudin, S. Osher, E. Fatemi, Nonlinear total variation based noise removal algorithms. Phys. D **60**, 259–268 (November 1992)
14. A. Chambolle, An algorithm for total variation minimization and applications. Journal of Mathematical Imaging and Vision **20**, 89–97 (2004)

15. T. Goldstein, S. Osher, The split Bregman method for l_1-regularized problems. SIAM J. Img. Sci. **2**, 323–343 (2009)
16. A. Marquina, Nonlinear inverse scale space methods for total variation blind deconvolution. SIAM J. Img. Sci. **2**, 64–83 (2009)
17. L. He, A. Marquina, S. Osher, Blind deconvolution using TV regularization and Bregman iteration. International Journal of Imaging Systems and Technology **15**, 74–83 (2005)
18. O. Michailovich, A. Tannenbaum, Blind deconvolution of medical ultrasound images: Parametric inverse filtering approach. IEEE Trans. Image Processing **16**(12), 3005–3019 (December 2007)
19. W.H. Richardson, Bayesian-based iterative method of image restoration. J. Opt. Soc. Am. A **62**(1), 55–59 (1972)
20. L.B. Lucy, An iterative technique for the rectification of observed distributions. Astron. J. **79**(6), 745–754 (1974)
21. P.A. Jansson, Deconvolution of Images and Spectra. Opt. Eng. **36**, 3224 (1997)
22. M. J. Cullum. Adaptive Optics. European Southern, Observatory, 1996
23. D. Dayton, B. Pierson, B. Spielbusch, J. Gonglewski, Atmospheric structure function measurements with a Shack-Hartmann wave-front sensor. Journal of Mathematical Imaging and Vision **20**, 89–97 (2004)
24. R.G. Lane, M. Tallon, Wave-front reconstruction using a Shack Hartmann sensor. Applied Optics **31**, 6902–6908 (1992)
25. R. Irwan, R.G. Lane, Analysis of optimal centroid estimation applied to Shack-Hartmann sensing. Applied Optics **38**(32), 6737–6743 (1999)
26. Y. Eldar, P. Kuppinger, H. Bölcskei, Block-sparse signals: Uncertainty relations and efficient recovery. IEEE Trans. Signal Process **58**(6), 3042–3054 (2010)
27. D.L. Fried, Statistics of a geometric representation of wavefront distortion. J. Opt. Soc. Am. **55**, 1427–1431 (1965)
28. V. Stanković, L. Stanković, and S. Cheng. Compressive image sampling with side information. In Proceedings of the 16th IEEE International Conference on Image Processing, ICIP'09, pages 3001–3004, 2009
29. T.O. Salmon, L.N. Thibos, A. Bradley, Comparison of the eyes wave-front aberration measured psychophysically and with the ShackHartmann wave-front sensor. Journal of the Optical Society of America A **15**, 2457–2465 (2007)
30. O. Michailovich, A. Tannenbaum, A fast approximation of smooth functions from samples of partial derivatives with application to phase unwrapping. Signal Processing **88**, 358–374 (2008)
31. M. Hosseini, O. Michailovich, *Derivative compressive sampling with application to phase unwrapping* (In Proceedings of EUSIPCO, Glasgow, UK, August, 2009)
32. I. Daubechies, *Ten Lectures on Wavelets* (SIAM, CBMS-NSF Reg. Conf. Series in Applied Math, 1992)
33. D. L. Donoho and Y. Tsaig. Fast solution of l_1-norm minimization problems when the solution may be sparse. Technical Report 2006–18, Stanford, 2006
34. S. Osher, M. Burger, D. Goldfarb, J. Xu, W. Yin, An iterative regularization method for total variation-based image restoration. Simul **4**, 460–489 (2005)
35. A. Savitzky, M.J.E. Golay, Smoothing and differentiation of data by simplified least squares procedures. Anal. Chem. **36**, 1627–1639 (1964)
36. A.N. Tikhonov, V.Y. Arsenin, *Solutions of Ill-Posed Problem*, vol. H (Winston, Washington, D.C., 1977)
37. Å. Björck, *Numerical methods for least squares problems* (SIAM, Philadelphia, 1996)
38. A. Beck, M. Teboulle, A fast iterative shrinkage-thresholding algorithm for linear inverse problems. SIAM Journal on Imaging Sciences **2**, 183–202 (2009)
39. O. Michailovich, An iterative shrinkage approach to total-variation image restoration. IEEE Trans. Image Process **20**(5), 1281–1299 (2011)
40. J.D. Schmidt, *Numerical Simulation of Optical Wave Propagation with Examples in MATLAB* (SPIE, Washington, 2010)

41. I. Daubchies, M. Defrise, C.D. Mol, An iterative thresholding algorithm for linear inverse problems with sparsity constraint. Comm. Pure Appl. Math. **75**, 1412–1457 (2009)
42. Z. Wang, A.C. Bovik, H.R. Sheikh, E.P. Simoncelli, Image quality assessment: From error visibility to structural similarity. IEEE Trans. Image Process **13**(4), 600–612 (2004)
43. Z. Wang, A.C. Bovik, Mean squared error: Love it or leave it? - A new look at signal fidelity measures. IEEE Signal Processing Magazine **26**(1), 98–117 (2009)

Chapter 5
Application: Surface Reconstruction in Gradient Field

Surface reconstruction from measurements of spatial gradient is an important computer vision problem with applications in photometric stereo and shape-from-shading. In the case of morphologically complex surfaces observed in the presence of shadowing and transparency artifacts, a relatively large number of gradient measurements may be required for accurate surface reconstruction. Consequently, due to hardware limitations of image acquisition devices, situations are possible in which the available sampling density might not be sufficiently high to allow for recovery of essential surface details. In this section the above problem is resolved by means of derivative compressed sensing (DCS). The results of this study are supported by a series of numerical experiments.

5.1 Derivative Compressed Sensing for Surface Recovery

The notions of photometric stereo (PS) and shape-from-shading (SFS) [1] are standard in computer vision, with their practical applications ranging from video surveillance to surface quality assessment. In both PS and SFS, a 3-D surface of interest is recovered from the measurements of its spatial gradient. In particular, under some reasonable assumptions on the light source and the object reflection properties, the unit normal to such a surface can be calculated from its grey-scale representation. Consequently, the normal can be used to recover its corresponding partial derivatives, followed by reconstructing an approximation of the original surface through the solution of a Poisson equation.

A practical difficulty in implementation of the above-mentioned techniques stems from the necessity to deal with relatively large sets of gradient data. Typically, such *dense* data sets are required to allow for accurate reconstruction of fine surface details, which are often occluded due to shadowing and transparency artifacts. In such cases, improving the acquisition requirements of the hardware in use through reducing the sampling density would unavoidably produce aliasing artifacts. Fortunately again we can overcome the above limitation, while allowing for accurately recovering digital signals from their sub-Nyquist measurements by means of compressive sensing.

M. Rostami, *Compressed Sensing with Side Information on the Feasible Region*,
SpringerBriefs in Electrical and Computer Engineering,
DOI: 10.1007/978-3-319-00366-5_5, © The Author(s) 2013

CS has already been used to tackle computer vision problems [2]. In this section, we introduce a method for reconstruction of 3-D surfaces from the sub-critical (incomplete) measurements of their spatial gradients.

Gradient space is the 2-D space of all (z_x, z_y) points. It is convenient to represent surface orientation in this space. In practice the gradient field is determined via the reflectance map $R(z_x, x_y)$ [3], which in turn is measured empirically. The reflectance map can be viewed as a 2-D image $i(x, y)$, where the image intensity is a function of z_x and z_y.

For Lambertian surfaces [3], the light is reflected in a given direction only based on the surface orientation. If the the measuring camera is placed at infinity (a single distant point source), the reflectance map based on Lambertian shading rule is given as [3],

$$R(z_x, z_y) = \frac{\rho(1 + z_x p_s + z_y q_s)}{\sqrt{1 + z_x^2 + z_y^2}\sqrt{1 + p_s^2 + q_y^s}} \tag{5.1}$$

where ρ is a reflectance factor.

The idea for both PS and SFS is to vary the viewing direction for measuring the x and y components of the gradient field of a surface, $z(x, y)$, at discrete points. Although the surface orientation is fixed, this will affect the reflectance map. For known ρ at least two views are required for determining z_x and z_y. But due to the nonlinearity in (5.1), more than one solution may exist. To emit such extra solutions, at least three measurements with three different light directions are required to solve uniquely for z_x and z_y. In practice, for improving the measurements, N images $i(x, y) = R(z_x, z_y)$ may be used ($N > 3$). These images result in the following equation for each point (x_i, x_j),

$$\begin{bmatrix} i_1(j, i) \\ \vdots \\ i_N(j, i) \end{bmatrix} = \begin{bmatrix} d_{1x} & d_{1y} & d_{1z} \\ \vdots & \vdots & \vdots \\ d_{nx} & d_{ny} & d_{nz} \end{bmatrix} \begin{bmatrix} \hat{n}_x \\ \hat{n}_y \\ \hat{n}_z \end{bmatrix} \tag{5.2}$$

where (d_{kx}, d_{ky}, d_{kz}) is the kth light ray direction and $\hat{\mathbf{n}}^T = [\hat{n}_x, \hat{n}_y, \hat{n}_z]^T$ is the surface normal vector. This equation in matrix form can be written as:

$$I = D\hat{\mathbf{n}}, \tag{5.3}$$

and the least square solution is given by

$$\hat{\mathbf{n}} = D^+ I, \tag{5.4}$$

where D^+ denotes Moore pseudo-inverse of D. Having the surface normal vector, the x and y components of the gradient field can be computed: $z_x = \hat{n}_x/\hat{n}_z$ and $z_y = \hat{n}_y/\hat{n}_z$. Consequently, over the whole surface the following measurements are obtained:

$$Z_x(j, i) = \frac{\partial z}{\partial x}|_{(x,y)=(x_i,y_j)}$$

$$Z_y(j, i) = \frac{\partial z}{\partial y}|_{(x,y)=(x_i,y_j)} \qquad (5.5)$$

For accurate surface reconstruction a high sampling density for the gradient field is required [3]. The sampling density is limited by the measuring device and there may be situations in which the sampling density is not sufficient for recovery of the surface details. This limitation may be resolved by applying DCS to this reconstruction problem. Having the partial measurements of matrices Z_x and Z_y, one can obtain vectors $\mathbf{b}_x$ and $\mathbf{b}_y$ via lexicographical column-stacking and similar to previous application use Algorithm 3 to solve for $\mathbf{z}_x$ and $\mathbf{z}_y$. Analogously this is equivalent with increasing the sampling density of the gradient field without improving the hardware device.

Algorithm 1: Derivative Compressive Sampling for Surface Reconstruction

---

1. *Data:* $\mathbf{b}_x, \mathbf{b}_y$, and $\lambda > 0$

2. *Initialization:* For a given transform matrix W and matrices/operators $\Psi_x, \Psi_y, D_x, D_y, T_x$ and T_y, preset the procedures of multiplication by A, A^T, B and B^T.

3. *Gradient field recovery:* Starting with an arbitrary $\mathbf{c}^{(0)}$ and $p^{(0)} = 0$, iterate (4.28) until convergence to result in an optimal $\mathbf{c}^*$. Use the estimated (full) partial derivatives $WT_x\mathbf{c}^*$ and $WT_y\mathbf{c}^*$ to recover the values of $\mathbf{z}_x$ and $\mathbf{z}_y$.

4. *Source recovery:* Use a Poisson solver to reconstruct the original source from its gradient field

---

Algorithm 3 summarizes DCS for surface reconstruction. In the final stage of Algorithm 3, it is required to solve a Poisson equation to yield the original source (the surface). Several approaches such as least square (LS) [4], algebraic [5], and l_1-minimization [6] have been proposed in the literature for this purpose. We use LS approach [4] in the current study for solving the Poisson equation.

5.2 Experimental Results

Simulated surfaces from [4] were used to assess the performance of the proposed method. The algorithm was tested over three surfaces known as Sphere, Peak-Valley, and Peak-Ramp. The surface lattices size is chosen 64×64, $\delta = 0.5$, and $\lambda = 0.001$. The subsampling matrices Ψ_x and Ψ_y were obtained from an identity matrix I through a random subsampling of its rows by a factor, r, resulting in a required partial sampling ratio. For sparse representation basis, again W was selected to be a four-level orthogonal wavelet transform using the nearly symmetric wavelets of Daubechies with five vanishing moments.

Table 5.1 Comparisons of surface recovery results

Image	Sphere				Peak-valley				Ramp-peak			
SNR (dB)	10	15	20	25	10	15	20	25	10	15	20	25
MSE comparison												
DS	0.0017	0.0017	0.0017	0.0017	0.0027	0.0013	0.0002	0.0001	0.0443	0.0139	0.0051	0.0033
CS	0.0057	0.0056	0.0055	0.0055	0.0210	0.0114	0.0103	0.0091	0.3773	0.2239	0.1201	0.0786
DCS	0.0022	0.0019	0.0018	0.0017	0.0071	0.0023	0.0006	0.0002	0.2464	0.0633	0.0157	0.0053

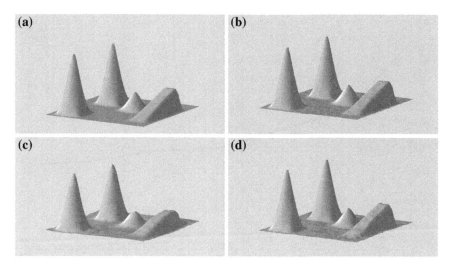

Fig. 5.1 Peak-Ramp surface (**a**) and its reconstructed versions using (**b**) DS, **c** classical CS, and **d** DCS for $SNR = 20\,\text{dB}$

For the purpose of comparison we have compared our algorithm with standard dense sampling (DS) and classical CS approaches in terms of MSE. The results of this comparison are summarized in Table 5.1 for different levels of noise and partial sampling ratio of $r = 0.5$ for classical CS and DCS. As expected, one can see that DCS results in substantially lower values of MSE as compared to classical CS, which implies a higher accuracy of surface reconstruction. As expected DS outperforms both methods but the performance of DCS is comparable and confirms the possibility of simplifying the hardware device using our approach without substantial reduction in reconstruction quality. The reconstruction result for Peak-Ramp surface is given in Fig. 5.1 for $SNR = 20\,\text{dB}$. Visual inspection on images, specially at the surface edges, confirms that DCS provides a result comparable with that of DS reconstruction. As it can be detected CS reconstruction results in smoothed edges in the ramp part of the surface, manifesting severe reduction of high frequency energy, which, by contrast, is well preserved in DCS reconstruction.

In another set of experiment we studied robustness of the proposed method towards noise addition. The cross-derivative constraints exploited by DCS effectively restricts the feasibility region for an optimal solution. Moreover, as explained in [7], the constraint $Bc' = 0$ in (4.27), can be considered as extra measurements of the sparse source. These measurements are noise free and consequently one can conclude that if we use this constraint, the reconstruction algorithm will become more robust towards the noise power. To investigate the robustness of the proposed algorithms towards measurement noises, its performances has been compared for a range of SNR values (as a measure for noise power) with classical CS. The results of this comparison are summarized in Fig. 5.2. As expected in both cases the reconstruction quality degrades with decreasing SNR, but this dependency is more critical for classical CS,

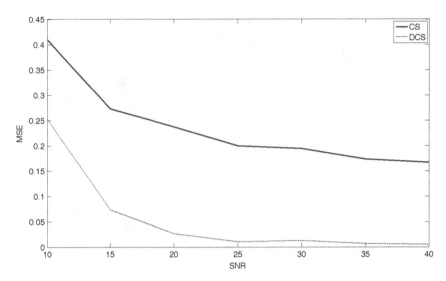

Fig. 5.2 MSE of surface reconstruction as a function of SNR. Here, the dashed and solid lines correspond to classic CS and DCS, respectively, and $r = 0.5$

which results in steeper graph in Fig. 5.2. This fact represents another advantage of incorporating the cross-derivative constraints in the process of surface recovery.

5.3 Summary

In this chapter, the applicability of the proposed scheme to the practical problem of surface reconstruction was demonstrated. To simplify the measuring devices, a DCS-based approach has been proposed. Experiments confirm the source estimates by DCS have better quality as compared to the case of classical CS and comparable as to the case of dense sampling. One direction for future work is applying the algorithm in designing the sampling devices for surface reconstruction. Applying the algorithm in the sampling device structure will improve the capability of reconstructing surface details in the presence of low density measurements.

References

1. H.D. Tagare, D. Hemant, R.J. DeFigueiredo, A theory of photometric stereo for a class of diffuse non-Lambertian surfaces. IEEE Trans. Pattern Anal. Mach. Intell. **13**, 133–152 (1991)
2. P. Boufouno, Compressive sensing for over-the-air ultrasound. In *Proceedings of the 2011 IEEE International Conference on Acoustics, Speech, and, Signal Processing*, pp. 5972–5975, (2011)

3. R.J. Woodham, Photometric method for determining surface orientation from multiple images. Opt. Eng. **19**(1), 191139–191139 (1980)
4. M. Harker, P. O'Leary. Least squares surface reconstruction from measured gradient fields. In *24th IEEE Conference on Computer Vision and, Pattern Recognition*, pp. 1–7, (2008)
5. A. Agrawal, R. Chellappa, R. Raskar, An algebraic approach to surface reconstruction from gradient fields. In *Proceedings of the tenth IEEE International Conference on Computer Vision*, pp. 174–181, (2005)
6. Z. Du, A. Robles-Kelly, F. Lu. Robust surface reconstruction from gradient field using the l_1 norm. In *Proceedings of the 9th Biennial Conference of the Australian Pattern Recognition Society on Digital Image Computing Techniques and Applications*, pp. 203–209, (2007)
7. M. Hosseini, O. Michailovich, Derivative compressive sampling with application to phase unwrapping. In *Proceedings of EUSIPCO* (Glasgow, UK, August, 2009)

Chapter 6
Application: Diffusion Fields Reconstruction Under Heat Equation Constraint

Reconstructing a diffusion field from spatiotemporal measurements is an important problem in engineering and physics with applications in temperature flow, pollution dispersion, and disease epidemic dynamics. In such applications, sensor networks are used as spatiotemporal sampling devices and a relatively large number of spatiotemporal measurements may be required for accurate source field reconstruction. Consequently, due to limitations on the number of nodes in the sensor networks as well as hardware limitations of each sensor, situations may arise where the available spatiotemporal sampling density does not allow for recovery of field details. In this chapter, the above limitation is resolved by means of using the proposed algorithm. We propose to exploit the intrinsic property of diffusive fields as side information to improve the reconstruction results of classic CS.

6.1 Introduction

Many natural phenomenon in physics are governed by diffusion equation, including temperature flow, pollution dispersion, and disease epidemic dynamics. In such applications, sensor networks are used as spatiotemporal sampling devices to sample and reconstruct diffusion fields [1]. In contrast to general multidimensional signals, the effect of temporal and spatial down-sampling are not homogeneous. Generally, it is more expensive to increase the spatial sampling density as more sensors are needed in the network, while temporal sampling density is only limited by each sensor hardware [2]. An efficient sampling scheme will have an impact on real world applications such as pollution detection [3] and plume source detection [4].

Inverse problems of the diffusive fields are generally ill-posed and require a relatively large number of measurements. Typically, such dense data sets are required to allow for accurate reconstruction of fine field details. In such cases, improving the acquisition requirements of the hardware in use through reducing the sampling density would unavoidably produce aliasing artifacts. To overcome this limitation,

M. Rostami, *Compressed Sensing with Side Information on the Feasible Region*,
SpringerBriefs in Electrical and Computer Engineering,
DOI: 10.1007/978-3-319-00366-5_6, © The Author(s) 2013

we apply the proposed algorithm for accurate reconstruction of sources from sub-Nyquist sampling rates.

In the current note, we consider spatiotemporal sampling and reconstruction of a 1-D diffusive field $u(x, t)$ governed by the heat equation:

$$\frac{\partial u(x, t)}{\partial t} = \gamma \frac{\partial^2 u(x, t)}{\partial x^2}, \quad t \geq 0,$$

$$u(x, 0) = f(x) \tag{6.1}$$

where γ is the diffusion coefficient, x denotes spatial domain variable, t denotes time domain variable, and $f(x)$ represents the initial field value.

If the initial field value is available, we can solve (6.1) for $u(x, t)$. However, in many situations, initial field value is not available [2], and it is not possible to derive $u(x, t)$ based on solely the partial differential equation constraint, as $u(x, t)$ varies dramatically with different initial condition. In these situations, we can measure spatiotemporal samples and use them to reconstruct $u(x, t)$.

Here, we take advantage of CS for efficient field sampling. It seems to be natural to reconstruct the source field using the fact that it satisfies the partial differential equation in (6.1) more efficiently. Specifically, we propose new CS formulation that incorporates the side information derived from (6.1) to improve the reconstruction quality of the standard CS, while resulting in substantial reduction in the required sampling density. We show that our efficient CS formulation can reduce the dimension of the feasible region in field reconstruction, resulting in better reconstruction quality.

6.2 Diffusive Compressive Sensing

Let $u(x, t)$ represents an original diffusive field which satisfies (6.1). For the sake of convenience, $u(x, t)$ is assumed to be defined over a finite-dimensional, uniform, rectangular lattice in $\mathbb{R}^2$. The discretized version of this field can be represented in a matrix $X \in \mathbb{R}^{N \times M}$. We assume that this field is sampled via a sensor network with N_s nodes which are deployed uniformly in the space and each sensor collects N_t uniform samples in time. Clearly, we have $m = N_t N_s$ measurements which can be represented in a matrix $Y \in \mathbb{R}^{N_s \times N_t}$ with $m = N_s N_t \leq NM = n$. X and Y can be concatenated into two column vectors $\mathbf{x}$ and $\mathbf{y}$ by means of lexicographic ordering, respectively. It is assumed that the observed version $\mathbf{y}$ of the vector $\mathbf{x}$ is obtained as $\mathbf{y} = \Psi \mathbf{x}$, where Ψ is a subsampling matrix which accounts for the effect of uniform downsampling. It is also assumed that $\mathbf{x}$ admits sparse representations with respect to a linear transformation W, $\mathbf{x} = W\mathbf{c}$. Finally, in order to apply CS to the problem, it is assumed that null(Φ) satisfies SSP by choosing Ψ and W properly.

Under the above conditions, CS-based reconstruction of the representation coefficients $\mathbf{c}$ can be performed according to (1.3). Our proposed diffusive CS algorithm uses the fact that $u(\cdot, \cdot)$ satisfies (6.1). Let D_x and D_t denote the matrices of discrete partial differences in the spatial and time directions, respectively. Then, the discretized version of the constraint (6.1) suggests that

$$D_t \mathbf{x} = \gamma D_x D_x \mathbf{x} \rightarrow (D_t - \gamma D_x D_x) W \mathbf{c} = 0. \tag{6.2}$$

Let $B := (D_t - \gamma D_x D_x) W$, $\Phi' = \begin{bmatrix} \Phi \\ B \end{bmatrix}$, $\mathbf{y}' = \begin{bmatrix} \mathbf{y} \\ 0 \end{bmatrix}$, and $\mathbf{n}' = \begin{bmatrix} \mathbf{n} \\ 0 \end{bmatrix}$, then:

$$\mathbf{y}' = \Phi' \mathbf{c} + \mathbf{n}'. \tag{6.3}$$

Algorithm 1: Diffusive Compressive Sampling

---

1. *Data:* $\mathbf{y}$, δ, γ and $\lambda > 0$

2. *Initialization:* For a given transform matrix W and matrices/operators Ψ, D_x, D_t, preset the procedures of multiplication by $A = \Psi W$, A^T, B and B^T.

3. *Diffusive field recovery:* Starting with an arbitrary $\mathbf{c}^{(0)}$ and $p^{(0)} = 0$, iterate (4.28) until convergence to result in an optimal $\mathbf{c}^*$.

   a. Use CS solver algorithm of [5] for solving the optimization problem in (4.28).
   b. Update the vector of Bregman variables $p^{(t)}$.

4. *Source recovery:* Use the estimated (full) sparse representation $\mathbf{c}^*$ to recover the values of $\mathbf{x} = W \mathbf{c}^*$.

Note that the problem (6.3) is an instance of the problem (3.3) and can be studied in the proposed CS framework. Algorithm 1 summarizes all the diffusive CS algorithmic steps.

6.3 Experimental Results

The proposed algorithm is tested over three different solutions of the heat equation (6.1) as the source field, denoted by $u_1(\cdot, \cdot)$, $u_2(\cdot, \cdot)$, and $u_3(\cdot, \cdot)$ for different boundary and initial conditions. The fields are assumed to be defined over the lattice $[0, 2\pi] \times [0, 1] \subset \mathbb{R}^2$, uniformly discretized with $M = N = 128 \rightarrow n = 16,384$. We set the boundary conditions to be non-homogeneous for $u_1(\cdot, \cdot)$ and $u_2(\cdot, \cdot)$, and homogenous Neumann condition for $u_3(\cdot, \cdot)$. The initial conditions are chosen to be $f_1(x) = x$, $f_2(x) = \delta(x - \pi)$ (local point source), and $f_3(x) = \Pi(0, \pi)$ for each case, respectively. The subsampling matrix Ψ is assumed to downsample the source field uniformly with downsampling d_t and d_s factor in time and spatial domains, respectively:

Table 6.1 PSNR comparisons of diffusion field recovery results for noise level of 10 dB

d_s	1	2	2	1	4	4	1	8	8	1	16	16
d_t	2	1	2	4	1	4	8	1	8	16	1	16
PoS	50%	50%	25%	25%	25%	6.25%	12.5%	12.5%	1.56%	6.25%	6.25%	0.39%
PSNR comparison (in dB) for $u_1(\cdot, \cdot)$												
CS	14.07	20.04	13.57	6.73	7.41	5.72	0.65	0.62	−0.07	−0.51	−0.58	−0.06
DCS	24.95	25.22	21.61	21.46	21.43	14.96	17.36	17.88	11.61	14.00	14.50	10.45
PSNR comparison (in dB) for $u_2(\cdot, \cdot)$												
CS	14.03	19.97	13.70	12.38	15.54	11.57	6.79	7.20	5.87	0.61	0.58	−0.08
DCS	25.10	25.17	21.54	23.07	23.31	17.59	21.31	21.70	14.92	17.28	18.06	11.61
PSNR comparison (in dB) for $u_3(\cdot, \cdot)$												
CS	16.71	19.63	14.16	16.04	13.04	10.91	0.13	0.03	−0.37	−0.58	−0.59	−0.07
DCS	21.87	21.52	18.87	18.78	18.33	12.82	15.32	14.93	9.80	12.33	11.88	8.64

$$Y(i, j) = X(d_s i, d_t j), \quad 1 \leq i \leq N_s, 1 \leq j \leq N_t \qquad (6.4)$$

For sparse representation basis, W was selected to be a four-level orthogonal wavelet transform using the nearly symmetric wavelets of Daubechies with five vanishing moments and $\delta = 0.5, \lambda = 0.001, \gamma = 1$.

For the purpose of comparison, we have compared our algorithm with classic CS approach in terms of reconstruction SNR. The results of this comparison are summarized in Tables 6.1 and 6.2 for different levels of noise and different percentage of the samples (PoS) in each table. In each table results for downsamling factors of 2, 4, 8, 16 in different directions are provided. As expected, for all cases one can see that DCS results in substantially high values of output SNR as compared to classic

Table 6.2 PSNR comparisons of diffusion field recovery results for noise level of 40 dB

d_s	1	2	2	1	4	4	1	8	8	1	16	16
d_t	2	1	2	4	1	4	8	1	8	16	1	16
PoS	50%	50%	25%	25%	25%	6.25%	12.5%	12.5%	1.56%	6.25%	6.25%	0.39%
PSNR comparison (in dB) for $u_1(\cdot, \cdot)$												
CS	14.95	23.45	14.19	6.88	7.56	6.01	0.70	0.61	−0.06	−0.51	−0.57	−0.06
DCS	25.22	25.28	21.66	21.27	21.54	14.97	17.41	17.96	11.63	14.00	14.53	10.49
PSNR comparison (in dB) for $u_2(\cdot, \cdot)$												
CS	14.60	23.48	14.13	6.91	7.57	6.01	−0.08	−0.51	−0.57	−0.43	−0.58	−0.06
DCS	25.21	25.28	21.60	21.27	21.55	14.98	13.96	11.61	10.53	9.44	10.35	6.14
PSNR comparison (in dB) for $u_3(\cdot, \cdot)$												
CS	17.41	20.33	14.46	16.05	13.20	10.91	−0.07	−0.07	−0.35	−0.60	−0.61	−0.07
DCS	21.92	21.53	18.79	18.78	18.23	12.87	15.41	15.01	9.83	12.30	11.87	8.50

CS, which implies a higher accuracy of field reconstruction. A close look on both tables reveals interesting results of the proposed algorithm. Note if we downsample a source field in one direction with the same downsampling factor, regardless of the direction, the resulting number of measurements are the same. Now consider those columns of tables with the same downsampling factor but different direction, e.g. first and second column, while for the case of classic CS the reconstruction quality differs in both tables, the quality of reconstruction for the case of DCS is similar. This can be explained through different correlations of the samples in different dimensions. From (6.2) one concludes that a field sample $X(i, j)$ is correlated with $X(i + 1, j)$ and $X(i + 2, j)$ in spatial domain while it is only correlated with $X(i, j + 1)$ in time domain. In other words dependency of the samples are not the same in time and spatial domain and it is harder to reconstruct the field when we lack time samples which is reflected in CS reconstruction results. In contrast, when we apply DCS these dependencies are considered as an additional data and thus the reconstruction quality is similar and is independent of downsampling direction. Generally, when we encounter insufficient spatial samples, oversampling in time domain is used to compensate [2]. Our result indicates that DCS can recover the source with less time samples which can be translated as energy saving in sensor nodes.

Another important result is on robustness of the proposed scheme towards insufficient samples. Consider a row in Tables 6.1 or 6.2, it can be seen that as the downsampling factor increases the reconstruction quality for classic CS degrades severely and for downsampling factors of 8 and 16 almost no information is recovered. While for the case of DCS, the algorithm is robust and even when we downsample a field with factor of 16 in both directions, using almost 0.4 % of the samples, it still can recover some information. For better comparison Fig. 6.1 depicts performances of CS and DCS algorithm for a range of downsampling factors with $d_t = 1, SNR = 40$ dB, and $u_2(\cdot, \cdot)$. It can be seen that for the case of CS, the reconstruction quality degrades sharply for downsampling factors greater than 4 while diffusive CS is robust towards downsampling. This can be explained by the constraint exploited by DCS. The constraint $B\mathbf{c}' = 0$ in (4.27), can be considered as extra measurements of the sparse source which can compensate for insufficient real measurements. This can explain while the difference between CS and diffusive CS is negligible for small downsampling factors, why it becomes considerable as the scaling factor increases. When we have enough information to recover the source then constraint (6.2) does not provide considerable information but when we lack enough information, this constraint becomes more important.

A comparison between the result of Tables 6.1 and 6.2 also reveals that although reconstruction quality degrades as the additive noise power of measurements increases but DCS seems more robust towards the noise. To investigate the robustness of the proposed algorithms towards measurement noises, its performances has been compared for a range of SNR values (as a measure for noise power) with classic CS for the case $d_t = 2, d_s = 2$ and $u_3(\cdot, \cdot)$. The results of this comparison are summarized in Fig. 6.2. As expected in both cases the reconstruction quality degrades by decreasing SNR, but this dependency is more critical for classic CS, which results in steeper graph in Fig. 6.2. Again, this can be explained by the constraint exploited by DCS

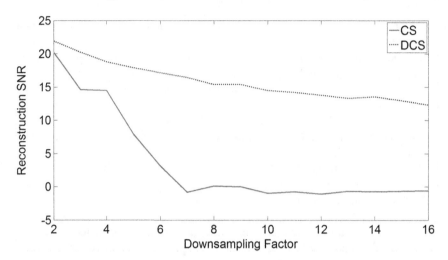

Fig. 6.1 SNR of field reconstruction as a function of spatial downsampling factor. Here, the *solid* and *dashed lines* correspond to classic CS and DCS, respectively, and $d_t = 1$

which restricts the feasibility region for an optimal solution. Moreover, as explained the constraint $B\mathbf{c}' = 0$ in(4.27), can be considered as extra measurements of the sparse source. These measurements are noise free and consequently one concludes that if we use this constraint, the reconstruction algorithm will become more robust towards the noise power. Intuitively one can say that since $\mathbf{n} \in \mathbb{R}^m$, $\mathbf{n}' \in \mathbb{R}^{n+m}$, and

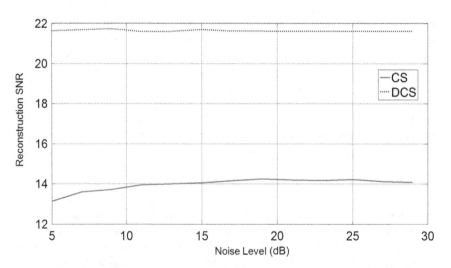

Fig. 6.2 SNR of field reconstruction as a function of noise SNR. Here, the *solid* and *dashed lines* correspond to classic CS and DCS, respectively, and $d_t = 2$, $d_s = 2$

$\|\mathbf{n}'\|_2 = \|\mathbf{n}\|_2$, the noise power has been multiplied by $\frac{m}{n+m} < 1$. This fact represents another advantage of incorporating the diffusive field constraints in the process of field recovery.

6.4 Summary

In this chapter, the problem of diffusive field reconstruction using sub-Nyquist sampling rates is studied. An efficient CS-based approach has been proposed to simplify the measuring devices and improve the device resolution. The proposed method applies CS for field reconstruction subject to an additional constraint, which stems from the intrinsic property of a diffusive field. Experiments confirm the source estimates by diffusive CS have better quality as compared to the case of classic CS and comparable as to the case of dense sampling. One direction for future work is applying the algorithm in designing the sampling devices for diffusive field reconstruction. Applying the algorithm in the sampling device structure will improve the capability of reconstructing diffusive field details in the presence of low density measurements. Another direction is to understand the performance under partial model knowledge.

References

1. P.C. Hansen, Rank-Deficient and Discrete Ill-Posed Problems: Numerical Aspects of Linear Inversion, vol. 4 (Society for Industrial Mathematics, 1987)
2. J. Ranieri, A. Chebira, Y.M. Lu, M. Vetterli, Sampling and reconstructing diffusion fields with localized sources, in *Proceedings of IEEE International Conference on Acoustics, Speech and Signal Processing* (IEEE, Prague, Czech, May 2011), pp. 4016–4019
3. A. El Badia, T. Ha-Duong, An inverse problem in heat equation and application to pollution problem. Inverse Ill Posed Probl. **10**(6), 585–600 (2002)
4. D.M. Moreira, T. Tirabassi, J.C. Carvalho, Plume dispersion simulation in low wind conditions in stable and convective boundary layers. Atmos. Environ. **39**(20), 3643–3650 (2005)
5. W. Yin, S. Osher, D. Goldfarb, J. Darbon, Bregman iterative algorithms for ℓ_1-minimization with applications to compressed sensing. SIAM J. Imaging Sci. **1**(1), 143–168 (2008)